SpringerBriefs in Environmental Science

SpringerBriefs in Environmental Science present concise summaries of cutting-edge research and practical applications across a wide spectrum of environmental fields, with fast turnaround time to publication. Featuring compact volumes of 50 to 125 pages, the series covers a range of content from professional to academic. Monographs of new material are considered for the SpringerBriefs in Environmental Science series.

Typical topics might include: a timely report of state-of-the-art analytical techniques, a bridge between new research results, as published in journal articles and a contextual literature review, a snapshot of a hot or emerging topic, an in-depth case study or technical example, a presentation of core concepts that students must understand in order to make independent contributions, best practices or protocols to be followed, a series of short case studies/debates highlighting a specific angle.

SpringerBriefs in Environmental Science allow authors to present their ideas and readers to absorb them with minimal time investment. Both solicited and unsolicited manuscripts are considered for publication.

More information about this series at http://www.springer.com/series/8868

Pamela E. Heckel

The Ethics of Energy Sustainability

An Energy Ethics Workbook

Springer

Pamela E. Heckel
The Five Minute Church, Inc.
Cincinnati
Ohio
USA

ISSN 2191-5547 ISSN 2191-5555 (electronic)
SpringerBriefs in Environmental Science
ISBN 978-94-017-9700-9 ISBN 978-94-017-9701-6 (eBook)
DOI 10.1007/978-94-017-9701-6

Library of Congress Control Number: 2015930842

Springer Dordrecht Heidelberg New York London

Printed on acid-free paper

Springer is part of Springer Science+Business Media (www.springer.com)

Preface

This book is an easy to use instructional aide. No two instructors will use it the same way! Explore sustainability issues in contemporary society through a transdisciplinary approach. Chapters include ethics, public resources, public policy, combustion, heat exchangers, nuclear, solar, water, and wind energy. A short summary is presented for each topic, followed by additional topics for research, assignments, and references. The complex assignments require students to grow in their professional judgment.

Challenging assignments lead to a broader understanding of the impact of engineering solutions in an economic, environmental, global, and societal context. Case studies are intended to be worked in groups and debated in class. Students are encouraged to research materials outside their disciplines to address problems they may encounter after graduation. Students are expected to rely on peer-reviewed scientific references for their opinions so that if challenged they can defend their positions.

Use interdisciplinary teams to resolve realistic case studies, and then debate these complex issues to determine which solutions are truly sustainable.Topics for debate were chosen specifically to allow the students to synthesize the information gleaned during their research. Discussions should include an analysis of the available raw materials, competing land uses, public policy, waste disposal, and energy demand.

It is assumed that each student is capable of performing a literature search. The answers to the assignments are not included in this text; some questions have no "right" answer. Many require students to become familiar with their local environment. Other questions are entirely open-ended. Students are expected to submit reports after conducting a review of the available literature. Students are encouraged to keep an open mind and to evaluate each topic objectively.

Science is never settled. The research of tomorrow may change our understanding of the world by unlocking new possibilities. New technology may transform the energy markets. May the quest never end!

Introduction

This text uses a transdisciplinary approach to encourage an informed discussion of "Energy Sustainability". The assignments are designed to increase awareness of the ethical challenges involved in setting energy policy. What exactly do we mean? Much of the current research on sustainability focuses on the carrying capacity of the land and the dangers of over population. This workbook focuses on the sustainability of energy resources which have applications: for transportation, for human comfort, for commerce, or for industry.

Sustainability

The term "sustainable" implies that the net negative effects of accessing and using a source of energy are zero. On the cost side are the efforts to extract and use the energy and to manage the waste stream. On the benefit side are the wages paid to workers, the jobs in rural areas, the value of the products made, and the improvement in the standard of living attributable to the energy source. A sustainable solution has a zero negative effect. There will be costs and benefits associated with the choice, but overall there is a positive outcome for humanity. The next few paragraphs provide clarification of this concept of sustainability.

The Earth contains different types of natural resources. Some like plants and water have many uses and are widely available. Other resources are sparsely distributed. The benefits of some, such as elements like uranium, are limited by our current technology. Energy-producing resources include flowing water, some plants, mineral deposits, some elements, wind, and the sun. Technology either exploits potential energy or harnesses kinetic energy for our benefit.

When the resource can be readily replaced or reused, it is renewable. Flowing water and solar power are renewable, in that rain falls and the sun shines. Organic fuels, involving plants such as sugar cane and corn which can be grown annually, are also considered to be renewable. Carbon-based fossil fuels, such as coal and oil, which depend on geologic processes, are not renewable.

Renewable fuels are not always sustainable. The rare earth elements used in solar panels are not widely available and must be handled carefully to avoid toxic effects. Hydroelectric dams cannot be built just anywhere. In addition to specific site requirements, settlements and cultural artifacts may be submerged by the reservoir behind the dam.

On the other hand, the extraction of non-renewable fossil fuels can have positive outcomes for some communities. Mining provides jobs, encourages settlements, and facilitates economic development. As the community expands, it attracts other businesses. If the cultural expectations include a balance of health, safety, environmental and economic concerns, the population will flourish. With access to reliable energy resources, the community will thrive. This is an economically sustainable scenario.

Unconventional fuels such as uranium may be sustainable for specific applications, but there is a finite amount of them. For example, nuclear-powered engines allow submarines to explore the seabed and plutonium is used to power space craft for interplanetary travel. Mining and enriching uranium for nuclear power is a hazardous venture. Significant environmental damage was caused by creating nuclear fuel at Hanford, the incident at Three Mile Island in 1979, the accident at Chernobyl in 1986, and the natural disaster at Fukushima in 2011.

Hydrogen gas has been used to power zeppelins. Fuel cells containing oxygen and hydrogen produce energy and water and have been proposed as an alternative transportation power source. Future technology may make hydrogen-powered vehicles affordable.

Energy

The Law of Conservation of Mass claims that mass can neither be created nor destroyed. Exothermic chemical processes release energy in the form of heat, but mass is conserved. Nuclear fission releases energy as atoms are split, but the mass of all the particles is equal to the original mass of the atom even though the original atom is changed. In the combustion process, a carbon-based fuel is converted to carbon dioxide (CO_2) and water, and energy is extracted. In a hydroelectric dam, potential energy is converted to kinetic energy that is used to turn turbines to create electricity. A solar panel converts light into electricity. Mass is conserved. Energy is extracted. Which of these processes is most beneficial to mankind? What are the costs and what are the benefits?

Plants sequester carbon through photosynthesis. The growth rate of trees and other plants is highly correlated with the ambient concentration of CO_2. As that concentration increases, yield per acre increases. The current concentration of CO_2 is less than 1 % of the air we breathe and poses no threat of suffocation for animals. Under the current "climate change" rules, plants grown and consumed as fuel within the same 12-month cycle are not counted in either the sequestration budget or emissions allowance. This accounting trick contributes to the support of "global warming" supporters for research into renewable fuels such as ethanol and biodiesel. Some "global warming" deniers support the same research for a different goal: energy independence.

Wood is the most abundant fuel on earth and if we were still hunters and gatherers, we would have plenty of it. People living in settlements, however, use wood for shelter as well as for fuel. They also cut down trees to make fields for crops.

Fossil fuels are geological deposits of carbon-based materials, generally assumed to have been plants and animals from an earlier era. The most abundant fossil fuel is coal. Coal is solid. Oil is a highly viscous liquid. Natural gas released from deposits of oil and coal may become trapped in underground rock formations. Fracking releases gas trapped in shale deposits. Each of these fuels burns and has been used for centuries to cook food, provide heat, or generate power. Petroleum products, including oil and natural gas, are also used as transportation fuels.

Methane gas (CH_4) is produced by the decomposition of rotting plant and animal materials. In the past, methane was flared at landfills. Today, CH_4 is collected and sold as a fuel. For animals, including humans, CH_4 is a by-product of the natural digestion process and is released in farts. I am waiting for an entrepreneur to figure out how to collect it from third grade boys.

The United States Environmental Protection Agency (USEPA) regulates emissions from combustion units. Through the combustion process, energy, water (H_2O), CO_2 and other compounds are formed. Pollution control devices can remove particulates and absorb acidic gases before the exhaust stream is released into the atmosphere. The individual species depends on the fuel type. For example, gaseous sulfuric acid and compounds containing fluorine (F), chlorine (Cl), bromine (Br) are created when coal is burned. Non-combustible materials such as silica remain in the ash.

Industrial Furnaces use heat to extract commercially valuable substances from ores, hazardous wastes, or other raw materials. Boilers produce a heated stream of liquid or gas. Incinerators destroy solid or hazardous wastes to reduce the volume. Each of these may be combined with an energy extraction device such a turbine. A cogeneration unit uses the steam produced by a boiler to drive a turbine. This cools the steam to a liquid. The hot water flows through pipes to provide radiant heat to nearby structures. The ground serves as a final heat sink and the water returns to the boiler inlet at the appropriate temperature.

Incinerators burn carbon-based materials at a variable rate under ambient operating conditions. Some incinerators are connected to boilers to produce steam to heat buildings. Some incinerators simply reduce the volume of solid waste. A medical incinerator burns medical wastes such as surgical sponges and bloody dressings, discarded human tissues, plastic tubing and other bio-hazards, to prevent the spread of disease. A sludge incinerator burns de-watered sludge at waste-water treatment facilities. A municipal incinerator may burn garbage, discarded furniture, yard wastes and other refuse.

A boiler operates under constant temperature and pressure using controlled flame combustion. It releases a stream of either a heated fluid or heated gases. For example, buildings may be heated by radiant heat that was emitted from hot water flowing through radiators connected through a system of pipes to a central boiler.

Fission also releases energy. Nuclear energy is released as particles are emitted from the nucleus of an atom. In a nuclear reactor, this process is a controlled reaction resulting in substantial amounts of electromagnetic radiation and kinetic energy. In nature, the process is called radioactive decay and occurs at a much slower rate. Through nuclear transformation, the substances remaining after a nuclear reaction are different elements with different chemical properties than the original substance.

The term "geothermal" implies that heat is extracted from the earth. Hot springs are places where water heated by natural processes underground is accessible from the surface. The concentration of minerals in these waters is typically higher than in sweet or potable water. Geothermal wells access the heated underground water and circulate it through a pipeline to nearby dwellings. Geothermal heat pump solutions rely on the constant temperature of the land several feet below the surface. Either a horizontal field is dug or a vertical shaft is bored and a series of pipes is laid. These connect to a heat exchanger inside a structure. Heat is added/extracted to maintain a comfortable temperature inside the building.

Capturing the power of the sun before those rays are reflected back into space is more difficult than it seems. Passive solar power involves the strategic collection of heat or light through the choice of absorbent materials, such as rock and water, and through the placement of architectural features such as windows and rooms. Landscape options, surface color and surface hardness determine the effectiveness of passive solar systems.

Solar radiation is absorbed by the earth. Thermal gradients within the upper and the surface boundary layers in our atmosphere affect weather patterns. The land heats up more quickly than the water, so during the day the wind blows inland. Thermal gradients in bodies of water contribute to storm intensity and location.

The temperature of the ocean is closely linked to its ability to absorb/desorb atmospheric gases such as CO_2 and its salinity. Ocean Thermal Energy Conversion (OTEC) systems use the difference between ocean surface temperatures and deep sea temperatures to drive a heat engine. The most effective application of OTEC is to desalinate water.

Active solar is the term used to describe the conversion of sunlight into electricity. As light enters a solar panel, an electrochemical reaction occurs. The resulting direct current can be stored in a battery or converted to alternating current for distribution along the power grid. Research continues to explore the use of organic materials, but for now, solar panels contain semiconductors made from hazardous materials.

The power of flowing water has been used for centuries, first for transportation along rivers and second to power machinery, such as a grist mill. Dams were constructed to limit the destructiveness of rivers and/or to create pools of water for drinking, washing, and recreation. Hydroelectric dams were built to convert the energy of falling water into electricity.

Wave power has attracted the attention of surfers and inventors. Devices have been built to take advantage of incoming and outgoing tides. The commercial success of these for long range power generation has not yet been proven, but they have been used successfully to power on-board devices.

Wind power is classified as a renewable resource because it depends on thermal gradients and the rotation of the earth, both of which are natural phenomena. Although sailing may be considered a recreational activity in some places, it remains a viable means of transportation and commercial activity in others. Wind mills convert wind energy to mechanical force. Local applications include pumping oil/water and grinding grain. Large scale wind mills use turbines to convert wind energy to electrical power.

In the United States, there are laws governing the acquisition and use of each of these energy resources. The Clean Air Act and the Clean Water Act regulate emissions and effluent. Acts of Congress control the use and disposal of federal lands, including the mineral rights. The law of eminent domain permits the taking of private property for just compensation for the common good. The US Supreme Court decided in *Susette Kelo, et al. v. City of New London, Connecticut* that private property could be taken by eminent domain for commercial development. The seized property is currently undeveloped.

Ethics

The study of ethics requires agreement on the common values underlying the social contract. If the cultural expectation favors work over leisure, then those endeavors which create opportunities for work will be given preference, business-friendly policies will be implemented, and energy resources will be allocated accordingly. The consequences of this preference must be factored into the costs associated with the cultural expectation, whatever it is. The morality or immorality of opinions, decisions and actions depends on the common values. If an action is illegal, it is also unethical. It is unethical to use deceit to obtain a contract. It is also unethical to substitute inferior materials to those required by contract specifications. Unethical situations may involve deceit, evasiveness, lying, substitution, over-statement, fraud, data manipulation, incorrect billing, etc. If someone's actions cause you to lose trust, then that person's actions may be unethical.

The study of the ethics of energy sustainability requires consideration of both the social contract between people living in groups and the availability of natural resources. This requires an understanding of a number of interrelated factors from a variety of disciplines. Decisions made when groups are small and isolated from one another are very different from policy for a global society. An examination of these decisions is warranted with a clear understanding that natural resources are not evenly distributed and cultural expectations vary. Energy sustainability is not a one size fits all proposition!

Since the dawn of the industrial revolution, demand for reliable energy has grown. Society has moved from burning wood and plant materials to coal to oil, gas and nuclear power. We harnessed the power of the rivers with dams and created devices to extract energy from the ocean waves and from the sun itself. Every innovation upset the status quo and changed how we live with one another. The social contract was renegotiated. We adapted. The study of the ethics of energy sustainability thus begins with this broad transdisciplinary picture.

Contents

Chapter 1
Public Policy and Public Resources

Abstract Nations enact domestic policies to allocate public resources, to generate revenue or to encourage specific behavior. Each rule benefits those who support it and creates a disadvantage for those who oppose it. Although a specific public policy may be issued from a central governmental office, all contexts are local. Natural resources are not uniformly distributed. Support for or opposition to public energy policy depends on availability of natural resources, capital, and transportation corridors.

The United States Energy Independence and Security Act of 2007 signed by President George Bush provided government support for the development of solar panels, wind farms, and other types of alternative energy with the goal of achieving energy independence from foreign sources of fossil fuels. It was designed to reduce waste, improve efficiency, and encourage innovation (USG 2007).

Keywords Keystone pipeline · Fracking · Horizontal drilling · Carbon offsets · Potable water · Renewable energy

1.1 Public Policy

Nations enact domestic policies to allocate public resources, to generate revenue or to encourage specific behavior. The United States Code of Federal Regulations contains the rules that interpret the laws. Each rule benefits those who support it and creates a disadvantage for those who oppose it. The rules are written by civil servants. Congress has the responsibility to review and approve regulations. These well-intentioned policies often have unintended consequences.

Consider the price of gasoline. Low gas prices may make a difference in the number of people who ride the bus to work. Low gas prices reduce the cost to ship goods, including food. Low gas prices make curbside recycling profitable. Low gas prices encourage travel for pleasure.

High gas prices may reduce the number of miles driven. Consider that in some sections of the Interstate Highway System, weathering of pavement, rather than wear from vehicles, is the reason for surface maintenance and pavement replacement. If fewer gasoline-powered cars and trucks are using the road, less tax revenue

P. E. Heckel, *The Ethics of Energy Sustainability,* SpringerBriefs in Environmental Science, DOI 10.1007/978-94-017-9701-6_1

is collected from the sale of gasoline, causing a shortage in the Federal Highway Trust Fund. This delays road maintenance and contributes to unemployment in the highway construction sector. Higher gasoline prices also increase the cost to grow, manufacture, and transport food.

Thus, a policy to double the tax on gasoline may initially generate more tax revenue to pay for road maintenance. It may also encourage more people to use public transportation or to purchase vehicles that use less gasoline. Over time, people will find some way to reduce their use of gasoline, so they can purchase other goods and services, which may cost more as a direct result of the higher tax on gasoline (Wright et al. 1997).

The fate of a Republic depends on the morals and ethics of its elected officials. In the United States, the offering of money to overlook a violation of regulations is considered a bribe (Wilcox and Theodore 1998). If a politician influences the passage of law favorable to a specific individual or company in exchange for campaign contributions or cash or gifts or other personal benefits, the politician may be censured for a violation of ethics by the legislative body. The practice of "quid pro quo", that is offering "this for that" generally refers to the exchange of services and/or money and/or political favors and is in many instances illegal and can result in incarceration and disgrace. When the responsibility for the enforcement of codes of conduct is assigned to a group of ethically challenged individuals, prosecution for ethics violations is apt to be tainted with favoritism, revenge, and political bias.

In the United States, it is not illegal for groups of individuals engaged in similar trade or craft or occupation to approach elected officials regarding legal protection of their livelihood. Unions are an example of this legal process to protect skilled craftsmen. Professional societies may request licensing of persons who desire to operate specific types of businesses. This is legal and includes engineers, doctors, lawyers, hair dressers, accountants, the food and beverage industries, etc. Many regulations are passed to protect the interests of consumers.

A contract is an agreement that binds two or more parties. In some communities, people reach agreements verbally and shake hands to seal the deal. Witnesses attest to the terms of the contract. A verbal contract is unenforceable and relies on the good faith and intentions of the parties. It is illegal to breach a written contract that has been witnesses by an impartial third party. Shunning is used in some communities to punish the party that fails to meet the terms of a verbal contract.

Substituting materials of inferior quality for those specified by contract is cheating and is unethical. The risk of getting caught varies. A defense contractor purchased circuit boards for several projects, one of which was a satellite. The supplier shipped recycled circuit boards rather than new. The satellite failed. The supplier went out of business.

Unethical situations may be associated with activities which are not illegal. Tolerance for these situations reflects personal and community values. For example, peer pressure is used to encourage recycling in the United States. Some communities provide bins to separate garbage from paper, glass, metals, and plastics and many of these communities may also have drop off centers for used electronics, paint, batteries, and other goods. Several charities collect used household items and resell them. Consumers of these used items come from every socioeconomic class. Persons who throw a glass bottle into the garbage bin may be reprimanded by observers who consider such behavior to be either thoughtless or immoral. In some communities, picking through the recycling bins for metals to sell is considered stealing, although picking through the trash is not.

Many immoral activities are neither illegal nor unethical. Laws may be passed to discourage sinful behavior, such as stealing and murder. The leaders of the Colonies believed that they were expected to lead by example. Skipping church was a sin. In the 1600s, community leaders attended church and punished anyone who did not. Laws prohibited commercial activity on the Sabbath, Sunday. As the nation grew and attracted persons who celebrated the Lord's Day on a different day, those pious persons celebrated their Sabbath and worked on Sunday. Eventually, these Blue Laws were challenged in court under the First Amendment and overturned.

Although a specific public policy may be issued from a central governmental office, all contexts are local. The economic engine of Appalachian mining communities is coal. The demand for coal drives all other development, from roads to transport it to schools and other services for the residents. The federal government owns much of the land. When the government passes regulations to discourage the use of coal, miners lose their jobs. Although government services are available in the cities, in rural communities, families lose their homes and go hungry. Thus, in Appalachia, the legal, anti-coal policies of the current Administration are considered both unethical and immoral.

In the United States, the rates that utilities can charge for electricity are set by a public utility commission. Proposed regulations include target utility rates for power plants using a mixture of renewable energy and combustion-based generation. US government policy favors the development of renewable energy. The profit margin is based on the expected cost for each kilowatt-hour (kwh). Solar and wind energy are more expensive per kwh than electricity from coal (USEPA 2014).

The Southern Company has demonstrated the feasibility of capturing, liquefying and storing CO_2 emissions. Some critics assert that the cost of CCS technology makes it unprofitable to build new coal-fired plants (Clean Techna 2014).

Public policy may encourage or discourage development. Strip mining in the Appalachian Mountains was made possible by the development of roads to rural and isolated communities. Previously inaccessible near-surface deposits were able to be harvested, contributing to local prosperity. Specific energy policies of the Obama administration were designed to reduce the use of coal and negatively impacted these communities. The USEPA's interpretation of the Clean Water Act required the treatment of waste water generated during the mining process and also required containment of storm water runoff to prevent flooding in sparsely populated areas. The new regulations were specifically crafted to reduce the profitability of strip mining. The policies led to widespread unemployment in coal mining communities.

Public energy policy in the United States is designed to reduce waste, improve efficiency, and encourage innovation. Here are a few examples of policies enacted under the Bush administration and continued under the Obama administration that will be presented in more detail in subsequent chapters of this text. The US

Congress passed the Energy Independence and Security Act of 2007 which called for investments in renewable energy and led to regulations requiring up to 10 % ethanol in every gallon of gasoline (USEPA 2010). The Department of Defense funds research to create alternative transportation fuels, such as biodiesel and organic jet fuel. Methane recovery systems on sanitary landfills capture methane emissions from decomposing garbage. Other policies encourage the use of solar panels, atria, geothermal heat pump solutions, windmills or other alternative energy sources to reduce lifecycle energy costs of new government buildings. There are reasons to support or oppose each of these energy choices depending on the location of the facility in relation to the available natural resources and transportation corridors.

1.2 Population

The ability of a given place to sustain life is closely linked to the carrying capacity of the land. This reflects the competing demands for food, water, shelter and fuel. If the demands placed on the land are sustainable, the population will thrive. We live with the tension between supply and demand, war and peace, life and death.

Sociology textbooks warn that the world population growth rate is unsustainable. The same prediction was made a hundred years ago. The "population bomb" supporters point to areas of the world where people are sick, starving or lack access to potable water. Some of this misery is linked to public policy and some of it reflects personal choices. Two distinctly different worldviews are now apparent: the culture of life and the culture of death.

The culture of life estimates that the entire world population, 7 billion in 2010, could live in the Republic of Texas in the United States. They celebrate the development of antibiotics such as penicillin and the anti-retroviral drugs. They lament a breakdown in family structure. They advocate personal responsibility, education, work, monogamy, charity, and delayed sexual gratification. They believe that the human species has stewardship over the Earth. The leaders of the Christian, Hindu, and Buddhist faiths advocate for the culture of life.

Their solution to the "population bomb" relies on the natural process of life and death. Natural phenomena such cholera kill millions annually. The evolution and transmutation of fatal bacteria and lethal viruses occurs naturally. As of 1995, a genetic variant of the plague bacterium lives in Madagascar. The Ebola virus spread from Nigeria to several other countries in 2014. Since 2013, a nasty anti-biotic resistant bacterium exists in India. Currently, the HIV/AIDS infection rate in Africa is more than 40 % and some forms of the virus are resistant to anti-retroviral drugs. Some strains of syphilis and gonorrhea are now incurable.

In contrast, the culture of death sees a world with limited resources. They value all animal species except *homo sapiens*. In California, farmers were prohibited from taking water to grow crops, to protect the habitat of the snail darter fish. Pessimistic politicians with this worldview pass public policy that pits one group against another. Such policies encourage dissent and contribute to social unrest. They encourage genocide by fanning the flames of nationalism. Politicians appeal to greed, envy, and gluttony, all of which are condemned by the world's religions. They advocate infanticide, euthanasia, and abortion on demand specifically to reduce population pressures. They tout universal access to birth control. This in turn encourages sexual promiscuity that spreads sexually transmitted diseases, which may result in sterility or death.

Is it unethical for leaders to advocate behavior that is immoral? Abortion has been practiced in every nation for thousands of years and until 1881 it was unrestricted in the United States. It was also quite dangerous, but then again so was childbirth. The controversy over abortion on demand began as a labor dispute between herbalists, midwives and physicians. In 1881, physicians in Connecticut persuaded the legislature to make abortion illegal except when performed by a physician. Physicians in other states lobbied to restrict the practice of abortion to physicians. Eventually, abortion was criminalized. The abortion debate in the United States was decided, but not settled, by a Supreme Court decision in 1973, Roe vs. Wade. Physicians continue to insist that abortion should be a legal, surgical procedure, governed by the medical profession. Religious persons continue to oppose abortion. In 2013, the Republic of Texas passed HB2 requiring abortion facilities to comply with the standards of hospital-style surgical centers. That provision was ruled unduly restrictive and thus unconstitutional by a Federal Judge and is headed for the US Appeals Court for the Fifth Circuit (National Abortion Federation 2014).

Manmade technology affects population. Television has been credited with reducing fertility rates. Modern pharmacology extends life. Many nations are supporting an aging population. Remember the unexplained collapse of the Mayan civilization? Who can predict if the human race will breed itself into extinction or if virulent diseases will intervene?

1.3 Land

When the land was sparsely populated, there was ample land, food, water, and fuel for individual families. As population increased, the demand for resources increased. Communities struggled with the allocation of those resources. In sparsely settled areas, each person was completely responsible for individual survival. In densely populated areas, residents experimented with the merits of different economic models.

Consider this classic example from more than two thousand years ago. Early Christians inspired by Jesus and His followers formed communities where they shared all their possessions with each other. The failure of some of these was documented in scriptures. Specifically, in 2 Thessalonians 3:6–12 Paul admonished the fledgling commune at Thessalonica against idleness with the pronouncement that those who will not work, shall not eat.

Human nature had not changed when the New World was settled. The decision to allocate food based on need rather than ability led to mass starvation in the Plymouth and Jamestown settlements with documented instances of cannibalism. A subsequent decision to allow private ownership of land resulted in an abundant harvest.

Settlers moved. Some establish new villages. The people living closely in town faced additional challenges: bad air and water contaminated with human and animal waste. Trade was established between groups. Over time, settlements grew and trade grew. A division of labor emerged. People living in town imported food and fuel and exported finished goods. People living on the farm discovered ways to increase yield. Each group dealt with waste, refuse, and excrement and learned to protect clean water from contaminants. The infrastructure evolved. Closed sewer systems replaced open trenches. Wells and cisterns were replaced by potable water distribution systems connected to indoor plumbing.

The industrial revolution changed lives. Mankind found time to observe, to study and to create. New technology was developed. The internal combustion engine in the automobile replaced the horse for transportation. Tailpipe emissions replaced the stench of manure. Electric lamps replaced candles and kerosene lanterns. The cooking hearth was replaced by the stove and oven. Central heating systems made the fireplace decorative rather than essential. Antibiotics and anti-viral drugs were developed. Some diseases were almost eliminated. Each of these improvements extended the life expectancy of the population and increased the value of land.

It is obvious from looking at the map that there is limited amount of land available for human habitation. Each parcel has an opportunity cost associated with its use. Do you use it to produce fuel or food? Do you cover it for housing or commerce? Do you flood it, drain it, or dig it? What value does it have as a park or cemetery? Land can over time be used for more than one productive activity, but some activities are mutually exclusive.

One of the most significant predictors of land use is land ownership. Consider a plot of flat, tillable land. An individual may place a dwelling on it and use a portion of it to grow plants, such as grass or tomatoes. A corporation may use it for office space, or for farming, or to extract a commodity such as water or minerals. A government may use it for office space or for the public good (park, road, airport, and cemetery).

The issue of ethics arises when the ownership of the land is the over-riding consideration for a project planned for public benefit or funded by public dollars. A well-connected Senator owned land on a river bank and stipulated that location for a new bridge to connect two existing state roads. Was this ethical?

Another predictor of land use is the location. The banks of rivers, lakes, and oceans may be used for habitation, farming, or commerce depending on geological characteristics, available transportation, and proximity to settlements. There are many natural harbors that are not used for commerce because they are located outside of shipping lanes. In contrast, Panama was chosen as the site for a canal across Latin America connecting the Atlantic and Pacific Oceans for economic, geographic, and geopolitical reasons. Hoover Dam was built to reduce flooding of the Colorado River, to provide drinking water, and to generate electricity. The geological characteristics of the land and the proximity to settlements determined the specific location of the dam.

The likelihood of mineral resources being extracted from a specific parcel of land increases with corporate interest. Does it matter who owns the land? No. The land may be owned by the government or individuals or a corporation. Potential profit provides the motivation. Access to capital, labor, transportation, and market demand are important variables in the profit equation. Strip mining is feasible when the coal deposits are close to the surface, but other factors determine whether or not a specific mineral deposit will be recovered. The price of oil is the main attraction for the development of the Canadian tar sands.

Given President Obama's persistent, personal denunciation of Charles G. and David H. Koch brothers, who own Koch Industries, Inc., opposition to the Keystone Pipeline to transport the tar sands oil to refineries in Texas appears to be politically motivated http://en.wikipedia.org/wiki/Political_activities_of_the_Koch_brothers. The five environmental impact statements conducted on the route conclude that a pipeline is the safest means to transport oil across land. Billionaire Tom Steyer pledged $ 10 million dollars through the NextGen Climate PAC to support candidates running for office in 2014 who oppose the pipeline. The US Supreme Court accepted a case based on a Nebraska pipeline citing law which may affect the proposed route (Bergin 2014). Is this simply politics as usual?

Who owns public land? Theoretically, public land, any expenses required to manage it and all profits associated with it, belong to the people. Land may be owned by townships, counties, cities, states, and the nation. When a lightening strike causes a forest fire in a national park, federal dollars are used to contain it. Lease agreements between the land owner and lumber companies may require the company to maintain access roads in exchange for the right to collect fallen trees. Farmers may lease prairie land to graze their cattle by paying rent to the lease holder. Drilling for oil in parts of Alaska may require Congressional approval, but oil revenues are shared by law with the native people who collectively own the land. Responsible management of land resources may generate revenue for the public coffers.

The Cliven Bundy situation called attention to the practice of grazing cattle on public land. When it became a state in 1864, the Federal government retained unappropriated public lands of Nevada. The Supreme Court had previously ruled in 1845 that all land within the borders of a state not personal property of individuals or appropriated for specific purposes belonged to the state. In the Bundy situation, the Federal government wanted to change the location of habitat set aside per the Endangered Species Act to permit the construction of a solar farm. The new habitat coincided with the land the Bundy family leased from the state for grazing cattle. There are a number of legal issues to resolve, the first is to identify the legal owner of the land (Nevada or the Federal Government) and the second is whether the doctrine of eminent domain trumps the Endangered Species Act. One of the ethical issues is that the son of Speaker of the Senate Harry Reid represented a Chinese company ENN Energy Group in the acquisition of land to build a solar project in Nevada. Was Speaker Reid has lobbying on behalf of ENN? http://www.thenewamerican.com/usnews/politics/item/18142-new-probe-confirms-harry-reid-s-long-history-of-corruption

Public policy may be used to protect land resources for the common good. The United States Resource Conservation and Recovery Act (RCRA) of 1976 regulates the disposal of solid waste on land. The most stringent requirements apply to hazardous waste generation, transportation, storage, disposal and treatment.

Household wastes are exempt, even though some of them may be hazardous. The Act is covered in 40 C.F.R. 238–282. Another law, the Comprehensive Environmental Response, Compensation, and Liability Act (CERCLA), was passed in 1980, to provide funds for the remediation of contaminated sites.

1.4 Food

Hunters and gatherers, farmers and fishermen, herbivores, carnivores, vegetarians and omnivores, everyone has to eat. Preferences for certain foods depend on the availability of those foods in specific neighborhoods. Climate and soil determine what can be grown and when it can be planted. Habitat limits the availability of game and fish.

Billions of people live in cities without access to crop land, domesticated animals or places to fish. Their diet is limited to those foods that can be transported and may consist of non-perishable items such as processed grains, sugar, and canned foods. The poor may lack cooking skills, access to cooking equipment, transportation to food stores, and/or cooking fuel. Wealthy residents have more choices. Their diet may include fruits, vegetables, dairy products, meat, fish, and other perishable items in addition to non-perishable foods.

Public policy has a significant impact on the availability of food (Folsom and Folsom 2014). Take for example the mandate to blend up to 10 % ethanol per gallon of gasoline in response to the Energy Independence and Security Act of 2007, Public Law 110–140 (2007). During spring, summer and fall, the blended gas reduces transportation-related ozone levels in the troposphere by producing less NOx per gallon of fuel burned. In the Great Plains, corn is grown specifically for ethanol production (USDoE 2013). Thus the public policy to use corn for fuel decreases the amount of corn available for food. This increases both the price of corn meal and the price of meat. Other crops can be used to produce ethanol (USEPA 2010). In Brazil, the sap of the sugar cane plant is used to produce sugar. The rest of the plant is used to produce ethanol. Waste plant material is dried and burned. Brewers grains are sold for livestock feed. Both corn and sugar cane require a long growing season, approximately 120 days. Is it ethical to use food for fuel?

Farming is not an easy profession. In addition to the physical demands of the work, farming requires a high tolerance for risk. The entire cycle of planting, growing and harvest depend on the weather. Price supports and crop insurance reduce the risk of bankruptcy due to crop failure. They provide a minimum level of income which the bank considers when the farmer asks for a loan to purchase seed, fertilizer or labor-saving equipment.

In the United States, the 2014 Farm Bill contains regulations and incentives for farmers. The latest bill provides a mechanism for organic and small farmers to obtain crop insurance (Hanson 2010). The bill funds education and training programs, provides resources to ensure food security, and funds the Supplemental Nutrition Assistance Program for the poor.

The practice of zoning identifies permissible uses for a parcel of land. Zoning restricts the location and size of residential, commercial, and industrial districts. Each district may have a different tax rate depending on the infrastructure and services provided. Taxes on farm land are generally lower than taxes on residential land. Local government may offer tax incentives for corporations to locate in particular areas. Administrators must zealously guard the public interest to avoid ethical conflicts when granting changes in the zoning designation for specific properties.

Taxes on transportation fuels may influence the mode used to haul crops to market. Railroads compete with barges for a share of the business of shipping commodities. Trucks are generally used to haul perishable items. Transportation costs are added to the price of the item. Thus, lobster caught and served in Maine should be fresher and less expensive than lobster caught in Maine and shipped to Tennessee.

1.5 Air

The air we breathe is a natural resource. Seventy-eight percent of our atmosphere is nitrogen. Twenty-one percent is oxygen. Oxygen is required for human life. Various gaseous compounds make up the rest. One of those, carbon dioxide (CO_2) is required for photosynthesis. Without atmospheric CO_2, all life on Earth will die. Over the past century, the concentration of atmospheric CO_2 has increased from 3 to 4 ppm.

Building codes dictate the minimum rate to replace indoor air with air from outside a building. A specific business may increase this air exchange rate to reduce odors, to prevent respiratory disease, and/or to meet business objectives. For example, a company that makes wooden shipping pallets was able to expand their business after they installed local exhaust fans along the saw tables to remove sawdust from the air. The Occupational Safety & Health Administration (OSHA) has identified permissible exposure limits for workplace exposures. https://www.osha.gov/dsg/topics/pel/. The American Conference of Governmental Industrial Hygienists (ACGIH®) uses Threshold Limit Values (TLVs®). http://www.acgih.org/TLV/. The National Institute for Occupational Safety and Health publishes a NIOSH Pocket Guide to Chemical Hazards http://www.cdc.gov/niosh/npg/. This contains recommended exposure limits for hundreds of chemicals.

Building HVAC systems monitor both temperature and humidity. During the winter, one of the concerns with the air exchange rate is the loss of heated, humid air as it is exhausted to the atmosphere and replaced with dry outside air at a cooler temperature. Warm, moist air is likely to contain dust, bacteria, viruses and mold and may contain volatile organic compounds (VOCs) from the interior furnishings. The colder replacement air must be heated. The need to add humidity depends on ambient conditions.

In the United States, public policy controls ambient air quality. The Clean Air Act (CAA) of 1963 established research funding to study the problem of air pollution in the United States. Amendments in 1990 to the CAA assign local, state, regional and federal authority for the regulation of chemical and combustion process emissions. National Ambient Air Quality Standards (NAAQS) (40 C.F.R. 50) set specific limits of ambient concentrations of six criteria air pollutants: carbon monoxide (CO), sulfur-based oxides (SOx), nitrogen-based oxides (NOx), ozone (O_3), lead (Pb), and particulates ($PM_{2.5}$ and PM_{10}). These pollutants are monitored at the local level. The U.S. Environmental Protection Agency (USEPA) sets the acceptable levels of ambient concentrations, enforces environmental law, and levies fines on polluters.

The combustion process releases carbon and hydrogen atoms and energy from fuel. A pure, clean burning fuel such as methane (CH_4) releases energy, CO_2 and water (H_2O). Under specific conditions, the nitrogen (N_2) in the air is unchanged, however, high temperature and pressure in gas turbines leads to the formation of nitrogen-based oxides (NOx) (Seinfeld and Pandis 2006). Biofuels and fossil fuels contain elements other than carbon and hydrogen and combustion of those fuels releases those trace elements, too. Sulfur-based oxide emissions (SOx) are released during combustion of oil and coal (Schobert 2013). The amount released depends on the concentration of sulfur in the fuel.

Mobile sources of air pollution include all transportation vehicles. The primary chemical species of concern are unburned hydrocarbons (HC), NOx, Pb (in areas where Pb is added to gasoline), and CO. Catalytic converters reduce tailpipe emissions of NOx and CO. The soot from high compression engines burning diesel fuel is a carbon-based compound with a high molecular weight (Nice and Bryant 2000).

Research shows that every gallon of gasoline contains trace amounts of mercury (Hg), a heavy metal. Gaseous Hg enters the atmosphere. Ionic species of Hg react with ionic species of sulfur (S), chlorine (Cl), methane (CH_4) or hydroxyl (OH^-) radicals (Heckel et al. 2013). Other mercuric compounds are less common. Both soot and particulate mercuric compounds precipitate from the atmosphere due to gravitational forces, falling to the ground within 1500 m of the source (Heckel and LeMasters 2011).

Regulated stationary sources of air pollution from combustion include power plants, boilers, industrial furnaces and incinerators. The common chemical species of interest are the six criteria air pollutants previously mentioned. Businesses, such as a dry cleaner or chemical plant, emitting hazardous air pollutants (HAPS) have lower limits and require a different type of permit. Home heating furnaces and cooking equipment have efficiency standards for operations rather than emissions limits. Restaurants may be required to obtain an emissions permit in some cities. Emissions from home fireplaces are not regulated, although fire safety building codes apply.

Pollution control equipment is intended to capture or convert all chemical species except CO_2 and H_2O. A cyclone contains a mechanical device to remove solids like ash from the emissions stream. An electrostatic precipitator (ESP) uses an electrical charge to attract ions. SOx can be removed by contact with a slurry containing calcium carbonate. The exhaust stream then passes through a bag house where solids are removed by mechanical filtering. The resulting compound can be used to make drywall. Up to seven percent of total plant output may be required to operate the pollution control equipment. These are large devices. Coal-fired power plants built before 1970 may lack the space required to install pollution control equipment.

Non-combustion energy sources do not emit CO_2 (Patel 2006). Nuclear, hydroelectric, wind, and solar energy have other problems, which are discussed in those sections of the text.

There are many ethics issues pertaining to air pollution. The most obvious are related to the permitting process. The Clean Air Act Amendments of 1990 specify limits for the criteria air pollutants. A region with air cleaner than the limits is in attainment of the standard. A non-attainment area has a measured concentration of one or more criteria pollutants in excess of the allowed levels. Another set of regulations applies to non-attainment areas and these regulations may disallow specific types of economic activity.

Air quality monitors are supposed to be placed at locations that experience typical and representative ambient air conditions. The Cleveland area of Ohio is known for its industrial base. Monitor locations were determined following established criteria. For many years the area was in attainment. Research conducted at a Cleveland hospital indicated a concentration of respiratory diseases in one of the suburbs. Residents petitioned their elected officials to install an air quality monitoring station in the suburb. The readings at that station exceeded the NAAQS and placed the entire region in non-attainment status. The culprit: the lake effect. The wind from Lake Erie blew pollution from industrialized areas of Cleveland into the eastern suburbs. How tempting it must have been to falsify that data! (Wilcox and Theodore 1998)

Once a non-attainment area is identified, any new source or any significant modifications of an existing source must install costly pollution control equipment. New sources who want to locate in those areas can hire an environmental consultant to estimate the emissions from their proposed facility using accepted models. Local departments of environmental compliance maintain a list of permitted sources. Occasionally, a permitted source decides to shut down the permitted equipment before the end of the permit period. They notify the local officials. A team composed of the entity that wishes to install the new source, the consultant, and an environmental lawyer may approach the permitted source regarding the option to purchase the unused credits. This process maintains existing air quality in a non-attainment area. If the actual emissions from the new facility exceed the credits purchased, the owner of the new facility may be forced to operate at less than full capacity or may be denied a permit to emit.

Reporting emissions data presents another potential for ethical conflicts and honest mistakes. Companies keep records of the fuel they purchase. From time to time, a sample of the fuel is tested against the purchase specifications. The chemical composition of the sample is considered representative of the batch. Plant operating parameters are set to burn the fuel as reported in the test report. Exceedances of permit levels are fined.

The Acid Rain program was established in part in response to fish kills in New England lakes. Emissions of sulfur-containing compounds from coal-fired power plants located along the Ohio River mixed with humidity to form a weak concentration of sulfuric acid inside rain drops (Seinfeld and Pandis 2006). Limestone is a buffering agent, so lakes lined with limestone were not affected by the acidic rain. Several lakes in the northeastern states have granite bedrock. Granite is not a buffering agent. Over time, the pH of those lakes dropped, causing changes in aquatic life. A pH below 5 is toxic to many fish (USEPA 2012).

One provision of the Acid Rain program was Cap-and-Trade. An emissions inventory was taken of industrial sources that emitted SOx. This became the baseline. Over time, the allowed levels of SOx decreased. Industries had three options: install pollution control equipment, buy low sulfur fuel, or shut down. A facility with excess credits was allowed to sell them. A similar program was imposed to reduce NOx levels.

The success of Cap-and-Trade to reduce SOx and NOx emissions was noted. Proponents of non-combustion power sources and biofuels proposed limiting CO_2 emissions through a similar program, marketing it as the solution to anthropogenic global warming (Lomborg 2001). The Chicago Climate Exchange (CCX) founded at the turn of the century to buy and sell carbon credits, closed in 2010 and was sold for half a billion dollars to a commodities trading consortium in 2011. Trans-global companies and universities purchased credits on CCX to offset a percentage of their carbon emissions. Alarmists predicted a tax on breathing, while skeptics joked about a tax on cow farts. In 2009, politicians dropped their support for Carbon Cap-and-Tax, as the program became derisively known by voters opposed to the scheme (FoxNews 2010). Undeterred, the Climate Action Reserve began working with the California Environmental Protection Agency Air Resources Board to develop and manage carbon offset programs in the State of California, which implemented a Cap-and-Trade program in 2012 (State of California 2012).

Predictions of global warming have lost some credibility in light of satellite data which not only showed no increase in the average global temperature 2000–2010 but also measured record high levels of sea ice in the southern hemisphere near Antarctica (Muller et al. 2013; Lomborg 2001). Anticipated extreme climate events predicted by twentieth century models failed to materialize (Curry 2014); however, shifting patterns in rainfall were noted. The media obligingly replaced the term “Global Warming” with “Climate Change”.

Many universities which measured their carbon footprints for a decade or longer as part of a national effort are reluctant to discontinue those efforts. Academics conducting government-funded research to reduce carbon emissions have a vested interest in keeping those research dollars flowing from the public coffers. The Obama Administration continues to support the development of non-combustion energy sources. Government policy remains anti-coal; the USEPA is proposing to require permitted sources to capture and store carbon emissions (USEPA 2013).

1.6 Water

Water is another public resource. Surface water includes fresh and salt water. The Great Lakes between the United States and Canada are the largest surface reservoir of fresh water in the world. Artesian wells access water trapped below the surface between two tectonic plates. Potable water is suitable for drinking. The water at the intersection of a fresh water source and a salt water sink has a lower salinity than the salt water (Cussler 2009). This water is "brackish".

The Florida Everglades provide fresh water for the southern half of the state of Florida. The Everglades empty into Florida Bay near Miami between the Gulf of Mexico and the Atlantic Ocean. The importance of this river of grass, marsh and cypress swamps was not fully appreciated until 1980 when deteriorating water quality, marked by salt water intrusion at the southern end, was noticed. Restoration efforts include destruction of canals, abandonment of farm land, and limitations on water demand.

The hydrologic cycle follows a single drop of water as it moves through the environment. This drop begins as a single molecule in the sky. It attaches to a cloud condensation nucleus. More molecules attach, to form a drop. Millions of drops cluster together in clouds and eventually are deposited on the surface as rain. In the ocean, diffusion occurs between the dilute surface water and concentrated deep sea water.

When it rains on land, some of the water soaks into the ground. This water nourishes plants and is released in gaseous form through photosynthesis. It may continue through the soil until it meets an impervious surface. Then it flows over or around that surface to a point at a lower elevation. Some of the water travels overland as storm water runoff. Eventually water accumulates at a lower elevation. It may become part of a lake, stream, spring, or river. These fresh waters may be used for irrigation, for commerce, by industry or by communities. Water may be purified for drinking water. Some of it evaporates, starting a new cycle (Heckel and Dombek 2009).

The ocean is not uniform. Differences in salinity and topology create currents. Thermal gradients in bodies of water affect both vertical and horizontal circulation patterns and contribute to storm intensity. The gravitational pull of the moon creates the tides. Oceanic circulation patterns may explain the recently measured increase in CO_2 in the atmosphere above the northern hemisphere (Garrison 2012).

The Federal Water Pollution Control Act (40 C.F.R. 100–149) regulates water quality in the United States. In 1972, amendments known as the Clean Water Act (CWA) established the National Pollutant Discharge Elimination System (NPDES) which defined acceptable levels of specific pollutants released from point sources into waterways. Permits are required for point source discharge. Storm water runoff after a precipitation event is considered a non-point source. It is more difficult to assess ownership of carried contaminants; however, monitoring of upstream and downstream concentrations has been used to apportion pollutants to specific agricultural operations.

Laws are passed to determine who has access to available water resources. In some states, crop yields are dependant upon irrigation and farmers must apply for permits to use flowing surface water. Environmentalists may sue on behalf of an aquatic species whose existence is threatened by declining water levels in the river downstream of the agricultural district. International law protects water rights for land owners, communities and species living downstream on a river that traverses national boundaries.

Water borne diseases like cholera are a global problem. Fresh water is required for human life. City water districts have an obligation to maintain a distribution network and a treatment facility. Residents expect their tap water to meet certain standards of purity. The Greater Cincinnati Water Works added a UV disinfection system to its chlorine disinfection line at the Richard Miller Treatment Plant to kill bacteria, protozoa and some viruses (City of Cincinnati 2013).

Water distribution systems need pumps and reservoirs to meet demand. Pumping water to homes and businesses takes energy. Companies often discover significant energy savings through water conservation projects.

Older cities often designed their waste water treatment systems to accommodate storm water runoff. As the system expanded to accommodate suburban areas, these combined sewers overflowed retention basins, spilling raw sewage in rivers. In the Los Angeles, California, basin, large concrete canals were constructed after the flood of 1938 to carry runoff to the sea (US 1941). In Cincinnati, Ohio, residents were forced to disconnect roof drains from the combined sewers (Hamilton County, Ohio 2001). Not only did this lessen the volume flowing through the waste water system, but it also reduced sewer back flows into basements of houses in low lying districts.

The Village of Greenhills Ohio adopted ordinances to prohibit dumping of used oil, paint and other household fluids into the storm drains (Hamilton County, Ohio 2009). This preventative action helped to restore the water quality in nearby recreational areas connected to the Mill Creek Watershed. During the winter, snow removal crews use low sodium options such as sand and beet juice mixtures on the streets adjacent to the park. Dumping unwanted chemicals into the water system is illegal and unethical.

Dams retain water. The difference in elevation between the top of the reservoir and foot of the dam may be suitable for power generation. A dam interrupts the supply of water downstream. The reservoir behind the dam may be used for recreational purposes. The water itself may become part of a potable water supply system. Lawsuits are common when new dams are proposed and when existing dams are breached or deliberately removed.

Native Americans have been successful is requiring dams to construct fish ladders through the locks. The fish ladder at the Ballard locks in Seattle, WA is a significant tourist attraction. People watch the salmon migration from salt to fresh water during the spawning season.

The southwestern United States experienced drought during the past decade as a mass of warm water flowed through the Artic Ocean and drew storms away from the southwestern region (Philander and Holton 2003). While not as severe as the Dust Bowl of the 1930s, comparisons were made. Water levels behind Hoover Dam decreased, reducing the flow of water in the Colorado River for thirsty farm land in the Imperial Valley in California. Many farms reduced the acreage under cultivation. The once mighty Colorado River was merely a trickle where it enters the Gulf of California. In consideration of the drought in 2008, the US Fish and Wildlife Service reduced the amount of water available to farmers in the Sacramento-San

Joaquin Delta to protect the Delta Smelt, a small feeder fish (US DoI 2008). In March 2014, the 9th Circuit Court of Appeals upheld the rule (US DoJ 2014).

Lack of precipitation leads to crop failure. Beef production has fallen in the Texas panhandle as rainfall over the past three years has been insufficient for growing pasture land. Wildfires ravaged areas experiencing humidity levels below 20%. During 2008–2013, a significant volume of warm water moved across northern Canada into the North Atlantic Ocean. Related changes in the jet stream affected humidity levels in the arid regions of Texas. Wildfires were a constant worry. During the summer of 2014, humidity levels approached 30% and the incidence of wildfires in Texas dropped. Long term forecasts predict normal precipitation levels through the fall. Beef production in Texas is expected to increase in spring 2015 if pasture land recovers (North Plains Groundwater Conservation District 2014).

In addition to fishing rights and consumption demands, ethical issues pertaining to water include permits to emit. The Clean Water Act protects water resources and requires industries to treat effluent. Industries which use water in their processes may discharge into surface water features if the discharge meets specific water quality standards. Less stringent requirements are imposed if the effluent is pre-treated to attain a neutral pH (between 6 and 7) and sent through the sewer. Water emitted through a pipe into a surface water feature is called a point source discharge. Permits are required. Samples taken upstream and downstream of a polluter are often provided as evidence.

Non-point sources of pollution include any substances in surface runoff. The most common is the mixture of oils, metals and non-metallic solids that wash off the roadways. Swales in medians and along shoulders may slow the flow and prevent unwanted substances from entering the watershed through creeks and streams. Storm drains on bridges divert chemicals away from the structure, but older drains empty into the water below.

Farm land is another non-point source of pollution. Excess chemical fertilizers and pesticides wash into nearby surface waters after a rain event. Manure ponds contain nitrogen-rich liquids and fecal bacteria. If breached, these liquids may destroy aquatic ecosystems. A dead zone in the Gulf of Mexico contains insufficient oxygen to sustain life due to nitrogen-nourished algae (USEPA 2014). Although it is extremely difficult to hide a manure pond breach, but over fertilizing is harder to pinpoint. Modern DNA analysis of e-coli bacteria can identify the worst offenders.

Air pollution is a third source of non-point source water pollution. Particulates in the atmosphere form cloud condensation nuclei. When it rains, hails or snows, not only do the particles trapped in the precipitation fall but particles suspended in the atmosphere can also be scavenged. Dust clinging to trees, buildings and other things is washed away. The precipitation either soaks into the ground or runs off impervious surfaces. Air pollution becomes water pollution.

The importance of access to fresh water cannot be overstated. Water is required for many industrial processes, including the generation of electricity. Water is used in some types of pollution control equipment. Our vehicles use it as a coolant. We humans need it. Insufficient potable water leads to dehydration and disease.

1.7 Additional topics for research

Alaskan Native Claims Settlement Act, Eisenhower Interstate Highway System, Eminent domain, US Energy Act of 2005 Federal Aviation Administration, Fire fighting, Forestry management, Keystone Pipeline, Property rights, United States Bureau of Indian Affairs, United States Clean Air Act, United States Clean Water Act, United States Constitution, United States Department of Agriculture, United States Department of Defense, United States Department of Energy, United States Department of the Interior, United States Environmental Protection Agency, United States Geological Survey, Census, Sustainability, Centers for Disease Control and Prevention Winnable Battles, UNICEF, Natural selection, Evolution, Creationism, Pollution prevention, Zoning, Bureau of Land Management, RCRA, CERCLA, Eminent domain, Mineral rights, Farm Bill, Bureau of Indian Affairs, United States Department of Agriculture, United States Department of the Interior, United States Environmental Protection Agency, Nevada territory, Northwest Territory, Louisiana Purchase, Seward's Folly, Organic farm movement, Sustainable farming, Pesticides, Herbicides, Farm equipment, Farm Bill, Crop rotation, Farm Tax Policy, Acid rain, NAAQS, CAA, Asthma, Point source emissions, Non-point source emissions, Kyoto Protocol, Montreal Protocol, Global warming, Climate change, Carbon sequestration, Solar flares, Southern Oscillation Index, El Nino, La Nina, Meteorology, Earth Systems Sciences, Toxics Release Inventory, Emissions inventory, Aurora borealis, Clean Water Act, NPDES, Point source emissions, Non-point source emissions, Water quality standards, Potable water, Tennessee Valley Authority, Hoover Dam, Bureau of Indian Affairs

1.8 Assignments

1. Mountain top removal is a cost effective method to remove near-surface coal deposits. In Kentucky and West Virginia, more than 450 mountains have been topped. The USEPA has denied strip mining permits to recover coal resources on public land in these states based on the Clean Water Act. Read the Clean Water Act. Which sections of the CWA apply to mining operations? Was the denial of permits justified?
2. Los Angeles County contains an international airport (LAX), a waste water treatment facility (Hyperion), and an oil refinery (El Segundo). There is a public beach at the end on one runway. The waste water treatment facility and the refinery are south, lying between the airport and Manhattan Beach. The deep water currents of the Pacific Ocean off shore run north, although wave action flows primarily west-to-east. Which governmental body has jurisdiction over water quality in Santa Monica Bay? Does the Clean Water Act apply to point source releases from these facilities?
3. Planes landing at the Los Angeles airport occasionally need to jettison fuel before landing. The fuel contains unburned hydrocarbons. What guidelines

control when and where the pilot can dump fuel? Are there provisions in the Clear Air Act that apply to this situation?

4. For St. Patrick's Day, the Chicago River is dyed green. Is a permit required? Which governmental agency has jurisdiction?
5. The US Constitution contains checks and balances to distribute and to limit the power of the Executive, Judicial, and Legislative branches. How can the Legislative branch limit the power of an Executive branch agency?
6. The USEPA issued a regulation that stipulates the use of ethanol in gasoline. Consider the entire process: growing and harvesting corn or sugar cane, the production of ethanol, transportation of corn/sugar cane/ethanol, blending ethanol into gasoline, distribution of the blended fuel to fueling stations, storage and burning the blended fuel. Which government agencies are involved at each step of the process?
7. Evergreen trees and other plants naturally emit volatile organic compounds (VOC). Plastics also emit VOCs. At high temperatures, in excess of 2800°F, nitrogen in the air reacts with oxygen to form NOx compounds. Ozone (O_3) is formed by a photochemical reaction between NOx and VOCs. In much of the United States, this occurs primarily between April and October due to the angle of inclination formed between the earth and the sun. The Clean Air Act National Ambient Air Quality Standards include O_3. Which agency monitors air pollution locally and issues a smog alert when levels of pollution exceed the allowable levels?
8. What is Leadership in Energy & Environmental Design (LEED)? Describe the categories of points available. What are they intended to accomplish?
9. What is the Environmental Protection Agency's Energy Star program? Which appliances are covered? Who makes those appliances?
10. In 2010, the American Institute of Architects (AIA) issued a 20/20 challenge to "Design for a New Decade". Describe the core ideas of the challenge. Which of these ideas have been realized in policy?
11. What are the three key accomplishments proposed on the Architecture 2030 challenge? Is there a scientific basis for their benchmarks? What are the societal benefits of issuing a challenge? How would your life be affected if the 2030 challenge were met?
12. Was Union Oil compensated when the environmental regulations halted mining of rare earth elements in the Mountain Pass District, San Bernardino County, California? What happened?
13. The proposed 2013 USEPA regulations will force some utility companies to either install carbon capture and storage (CCS) devices or to shut down coal-fired power plants. The justification for the proposed regulations is the increase in CO_2 concentration over the past decade from 350 ppm to 700 ppm. The USEPA claims that the increase is due to anthropogenic activities. The "global warming" skeptics point to a mass of warmer waters which recently flowed through the Arctic Ocean into the north Atlantic Ocean, due to the natural circulation of the ocean. Profit is the motive. A well-connected utility which plans to build additional nuclear powered facilities is trying to sell power in areas

dominated by power plants burning fossil fuels. The company backs the global-warming-due-to-anthropogenic-activities scenario because nuclear energy does not emit CO_2. Is this ethical?

14. CASE STUDY What was the motivation for the USEPA ethanol mandate? Has the mandate had the intended effect? What are the economic consequences? What are the societal consequences? What are the safety concerns? If the policy were repealed, what might happen? Use the worksheet Fig. 1.1 to organize your arguments. Is the USEPA ethanol mandate ethical? Is it sustainable?
15. For each of the G20 countries, determine the dominate energy source and the estimated reserves. For example, in the United States, 40 % of our electricity comes from coal and our reserves contain a 200 year supply. [These numbers may have changed.]
16. What is the population density of each G20 country and the per capita annual energy usage? Use graphs and tables to compare the energy demands. What factors are most likely to influence energy demand?
17. On a map, identify the areas of armed conflict around the globe. What economic factors motivate the war? What other factors are used to recruit troops?
18. On a map of your country, identify locations with the highest and lowest population density. What makes some locations more attractive than others?
19. Why do people leave the country of their birth to migrate to other lands? Which countries are the most/least welcoming for immigrants? Which countries have the highest/lowest immigration rates? [Hint: the answers will not be the same.]
20. Is there a relationship between immigration and the average minimum wage for labor?

USEPA 10% Ethanol Mandate

Health & Safety	Environment	Economy
Pro:	Pro:	Pro:
Con:	Con:	Con:

Fig. 1.1 USEPA 10% Ethanol Mandate

21. Gather data to compare the system of governments of the members of the United Nations. Is there any significant correlation between the standard of living and the system of government?
22. Using World Health Organization data, make a table to identify countries in Africa with HIV infection rates in excess of 15 %. Are these the same countries where consensual same-sex relationships are legal? How is the virus affecting the social fabric of these countries?
23. Why did Uganda pass a law making "acts of aggravated homosexuality" illegal in 2013?
24. The worldwide ban on DDT allowed mosquitoes to become the most effective disease vectors ever known, infecting millions with malaria, typhoid, yellow fever and other tropical diseases. Was the ban on DDT ethical?
25. In the United Kingdom, abortion is legal. The procedure is covered by national health insurance. Aborted fetuses are not given to the patient for burial; they are discarded as medical waste. One hospital incinerates all medical wastes and uses the energy to heat/cool the hospital. Is this ethical?
26. CASE STUDY. Senator X proposes a law to reduce the number of teen mothers in State Y. Under this law, babies born to teen mothers would immediately after birth become wards of the state. The age of the father is irrelevant. The economic circumstances of the teen herself or her family is irrelevant. Children who are wards of the State live in either orphanages or foster homes and are immediately available for adoption. Wards of the State are given access to free medical care and preschool education, and funds are provided for their food, clothing and material necessities. Identify who would benefit from this law. What are the likely unintended consequences? Will this reduce or increase the incidence of teen pregnancy? Is this an ethical solution to control population? Is it sustainable? Organize your facts on the worksheet, Fig. 1.2. Debate this proposal.
27. One of the most toxic of wastes is radioactive waste in the form of spent fuel rods from nuclear reactors. After extensive geological assessment, the caverns of Yucca Mountain were designated in 1987 to be a repository for nuclear wastes. This mountain lies approximately 100 miles from Las Vegas, Nevada, on Federal land. In 2010, the Obama Administration abandoned the project for political reasons. Was this ethical?
28. Ideally, lechate from a landfill does not contaminate the ground water. It is easier to collect and treat landfills built on bedrock or impervious soils than on alluvial or sandy sails. Identify areas in your country/state where a leaky landfill would not be a problem.
29. The Cherokee Nation used to lie in what is now part of the states of Georgia and North Carolina in the United States. Contrary to an interpretation of treaty law by the US Supreme Court, President Andrew Jackson allowed the removal of the Cherokee by force, the confiscation and redistribution of their lands by lottery, and resettlement in Oklahoma. Compare the typography of the native Cherokee homestead with the topography of the lands they were given. What cultural changes may have occurred within the Cherokee Nation that could be attributed to a change in their environment?

Children of Teen Mothers Become Wards of the State

Health & Safety	Environment	Economy
Pro:	Pro:	Pro:
Con:	Con:	Con:

Fig. 1.2 Children of teen mothers become wards of the state

30. CASE STUDY "Lithium supply security has become a top priority for Asian technology companies." http://minerals.usgs.gov/minerals/pubs/commodity/lithium/mcs-2014-lithi.pdf "In 2009, the U.S. Department of Energy awarded the company $ 9.5 million to construct the first U.S. recycling facility for lithium-ion batteries, which was expected to be operational by year end 2013." What is the status of this facility? Is recycling a sustainable alternative for Lithium batteries? Under what conditions should the recycling of Lithium batteries be mandatory? Use the worksheet Fig. 1.3 to organize your ideas.
31. The California Gold Rush and the Alaskan Gold Rush attracted prospectors. Who became rich?
32. What is eminent domain? Summarize the Supreme Court's decision in Kelo v. City of New London, 545 U.S. 469 (2005). How is that property being used today? How was Norwood, Ohio v. Horney 110 Ohio St.3d 353 different? How is that property being used today?
33. Why was the Eisenhower Interstate Highway System built? How were land parcels chosen? How were neighborhoods affected?
34. The Keystone Pipeline was approved in several studies as the most economical and environmentally safe means to transport oil from Canada to refineries in the United States. What were the alternatives for the Canadians? Why was the Keystone Pipeline delayed? Was that decision ethical? Would the Keystone Pipeline be sustainable?
35. During 2009–2013, the average income in the United States dropped, the work force participation rate dropped, the unemployment rate dropped, and the price

Mandatory Lithium Battery Recycling

Health & Safety	Environment	Economy
Pro:	Pro:	Pro:
Con:	Con:	Con:

Fig. 1.3 Mandatory lithium battery recycling

of farm land increased. Identify possible demographic trends and/or public policy behind each of these reported statistics.

36. What resources make the nations of Georgia and Ukraine attractive to Russia? Why is Crimea attractive? What was Russia's justification for annexing Crimea in 2014?
37. 4. Africa is known for its rich deposits of diamonds and gold. In the Central African Republic, a Christian-majority nation, most of the mining is done by Christians and most of the trading is done by the Muslim minority. In 2013, there was a coup led by the Muslims. Did the religion of the new (Muslim) government have any bearing on the diamond industry? Explain. Compare this situation with the sectarian violence between various sects of Muslims in the Middle East and/or past Catholic/Protestant violence in Northern Ireland.
38. CASE STUDY. What is the purpose of the North Atlantic Treaty Organization (NATO)? Who belongs to NATO? Does NATO have a strategic interest in Crimea? Use the worksheet Fig. 1.4 to develop a list of pros and cons for NATO military intervention to oppose the Russian annexation of Crimea.
39. How and why does the price of petroleum affect the cost of food?
40. Did the Fukushima disaster affect the quality or quantity of food in Japan? What is solar-sharing?
41. What was the effect of the USEPA ethanol mandate (up to 10% per gallon of gas) on the price of corn? Is it ethical to convert food crops to fuel? Explain.
42. The USEPA issued a ruling to protect the Delta Smelt, Hypomesus transpacificus, population. How did this ruling affect California's Central Valley farming community? Was this policy ethical?

NATO Opposition to Russian Annexation of Crimea

Health & Safety	Environment	Economy
Pro:	Pro:	Pro:
Con:	Con:	Con:

Fig. 1.4 NATO Opposition to Russian Annexation of Crimea

43. What is the relationship between an El Nino event and precipitation along the western edge of the continental United States? Joe Bastardi (2014) "The Coming Nine Months: Mud in the Eye of the So-Called 'Permanent' Drought?" https://patriotpost.us/opinion/25957
44. Is there a difference between the growth rate of urban and rural trees of the same species? What factors contribute to this phenomenon?
45. Soft wood trees such as pine grow faster than hardwood trees such as oak. Which species sequesters more carbon annually? Demonstrate by graphing tree growth rates over time.
46. In which zones is it possible to make a living by farming as little as five acres? For a given geographical location, identify potential revenue sources, crop selection and rotation practices, and livestock.
47. What is a (tree) wind break and how is it different from a snow fence?
48. Is radiation in the water from Fukushima a problem for commercial or recreational fishermen in the Pacific Ocean? Justify your answer.
49. Consider an abandoned but structurally sound 3-story parking garage that covers an entire city block. How could this be re-purposed to grow food?
50. CASE STUDY. In the United States, more than two centuries ago, sugar growers petitioned Congress to limit the allotment of land to grow sugar cane. In addition, the Federal Farm Bill sets the minimum price for cane sugar sold in the United States. The price for corn syrup and cane sugar are approximately the same due to the USEPA ethanol mandate. Both corn and sugar cane can be used to make ethanol. The price for cane sugar outside the United States is significantly less. You work for a drink manufacturer. A taste test indicates that consumers prefer drinks made with sugar instead of syrup. Would it be a good

business strategy to switch from corn syrup to cane sugar? What information do you need to make a decision? Use Fig. 1.5a, b to list the pros and cons of each sweetener from a sustainability perspective. How would the company persuade consumers to try the new formulation? Develop a marketing slogan for the national television audience.

51. Roof top gardens are recommended as a solution for urban hunger. What types of food can be grown? What is the difference between aquaponics and hydroponics? What safety features are recommended for roof top gardens? NIOSH (2013) Architectural Design and Construction—Instructor's Manual http://www.cdc.gov/niosh/docs/2013-133/
52. A "food desert" is a residential area which lacks access to grocery stores, farmers' markets or other places to purchase food. Typically these are in urban areas. Identify a "food desert" in your locality, if any. What challenges must be considered before the situation can be remedied?
53. How is ozone formed? When is the ozone "season" at your latitude?
54. AERMOD and CalPuff are two air pollution dispersion models. When would you use one or the other?
55. Why were Chlorofluorocarbons limited by the Montreal Protocol?
56. What is photosynthesis? What happens to crop yields as the concentration of CO_2 increases?
57. What is carbon capture and storage (CCS) and is it economically sustainable? Has the technology been developed to capture CO_2 from power plant stacks? Is it commercially viable?
58. Name three industrial applications for liquid/gaseous CO_2.
59. During Lisa Jackson's term as Director of the USEPA, humans were exposed under experimental, laboratory conditions to elevated levels of diesel emissions. Was this research ethical? List the criteria for ethical research involving humans. When were these research protocols developed?
60. Compare the urban ambient air quality in Pittsburgh, PA during the late 1960s to their air quality today. What factors contributed to the change?
61. What is a thermal inversion? What pollution event occurred in Donora, PA in 1948? What was the Great Smog of 1952 in London, England? How did these two events influence the environmental movement?
62. What are the limits established for criteria air pollutants covered by the National Ambient Air Quality Standards? Identify three non-attainment areas for each of the criteria air pollutants. Describe air pollution control devices invented to manage each of these pollutants.
63. Asthma research focused on the concentration of diesel particulates in the breathing zone of children formed a compelling scientific basis for a school bus retrofit program that has been adopted across the country. Develop a presentation on this topic. Include relevant statistics, such as rates of adoption, and summarize the current research on asthma. Has this been an effective intervention?
64. A nail salon plans to occupy a space in a shopping mall in your town. What are the applicable indoor air quality standards? Compare the OSHA permissible exposure limits, https://www.osha.gov/dsg/topics/pel/, the ACGIH Threshold

Cane Sugar in Soft Drinks

Health & Safety	Environment	Economy
Pro:	Pro:	Pro:
Con:	Con:	Con:

a

Corn Syrup in Soft Drinks

Health & Safety	Environment	Economy
Pro:	Pro:	Pro:
Con:	Con:	Con:

b

Fig. 1.5 **a** Cane Sugar in Soft Drinks. **b** Corn Syrup in Soft Drinks

Limit Values (TLVs®), http://www.acgih.org/TLV/, and the NIOSH recommended exposure limits http://www.cdc.gov/niosh/npg/ for acetone. Which standard is more conservative? Which standard is the nail salon required to meet? Which standard is more attractive for patrons?

65. What causes sick building syndrome? What steps are required to fix the problem?
66. Smoking is permitted in the Last Call Bar. The bar accommodates 100 patrons and 5 staff. Fire Codes require 36 sq. ft. of floor space per occupant. How

much floor space does the bar need? The ceilings are 20 ft. high. What is the volume of this space? On average, each patron smokes one cigarette each hour. Each cigarette emits 14 mg of particles. State your assumptions. Calculate the mass of cigarette-related particles in the air after 1 h of operation. Thirty air exchanges per hour are required to meet code. http://www.engineeringtoolbox.com/air-change-rate-room-d_867.html. What flow rate Q must the HVAC system supply?

67. Urban air quality has improved over the past 40 years. Why has the incidence of asthma increased over the same period?
68. What is "epigenetics"? Give an example.
69. Explain the Southern Oscillation Index. How does water temperature in the South Pacific influence water temperature in the Northern Hemisphere?
70. Why do ocean currents (and toilets) flow counter-clockwise in the Northern Hemisphere and clockwise in the Southern Hemisphere?
71. Why is the 1990 flood in Shadyside, OH considered a significant hydrologic event? How does it compare to annual floods along the Ganges River delta in Bangladesh?
72. South of the Mississippi River delta near New Orleans, LA there is a dead zone in the Gulf of Mexico; the oxygen content of the water is too low to support marine life. Name three other marine areas with dead zones. What caused the dead zone to form?
73. Some marine biologists speculate that eutrophication is to blame for dead zones. What evidence supports this theory?
74. During the second week of July 2013, thousands of dead fish littered the public beaches of Ludington, MI. Residents appeared unconcerned; families were playing in the water and sun bathing on the beach. Did the dead fish constitute a public health hazard? Propose a scenario for removing the fish from the beach.
75. Imagine you are the mayor of a costal town. A red algae bloom offshore killed local marine life and turned the water red. A local religious leader used the occasion to preach about the "End Times" prophesy. People are panicking. Is the water safe to drink? Should the beaches remain open? How do you calm the residents? Develop a message map for this scenario.
76. Explain the increase in estrogen concentration in our nation's river water. What is the effect on marine life? Are we being exposed to estrogen through our drinking water? Describe an available technology to remove estrogen and other trace elements from water. Is this process being used to treat the tap water you drink?
77. A homeowner has a cistern. Rain falls on the roof and is collected in the cistern. How can the homeowner determine if this water is safe to drink? The roof needs to be replaced. Should the homeowner select roofing materials that contain copper sulfate? Why is there copper sulfate in the roofing materials?
78. During the summer of 2014 the water supply for the City of Toledo OH was affected by a toxic blue-green algae bloom in Lake Erie. What was the cause of this algae bloom? What steps can be taken to reduce the likelihood of another incident?

References

Begin N (2014-04-08) Nebraska lawsuit delays decision on Keystone XL. Lincoln Journal Star Online. Accessed 14 Aug 2014

City of Cincinnati OH (2013) Treating water with ultraviolet light. http://www.cincinnati-oh.gov/water/about-greater-cincinnati-water-works/major-projects/uv-treatment-project/. Accessed 30 Sept 2014

Clean Techna (2014) Coming coal & nuclear power plant retirements, & what that means. http://cleantechnica.com/2014/05/05/power-plant/. Accessed 30 Sept 2014

Curry J (2014) Climate science: uncertain temperature trend. Nature Geosci 7.2:83–84

Cussler EL (2009) Diffusion: mass transfer in fluid systems, 3rd edn. Cambridge University Press; ISBN-10: 0521871212; ISBN-13: 978-0521871211

Folsom BW Jr, Folsom A (2014) Uncle Sam can't count: a history of failed government investments, from Beaver Pelts to Green Energy. Broadside Books. ISBN-10: 0062292692; ISBN-13: 978-0062292698

Fox News (2010) Collapse in Chicago climate exchange means strategy shift for global warming curbs. http://www.foxnews.com/politics/2010/11/09/collapse-chicago-climate-exchange-means-strategy-shift-global-warming-curbs/. Accessed 30 Sept 2014

Garrison TS (2012) Oceanography: an invitation to marine science, 8th edn. Cengage Learning. ISBN-13: 978-1111990848; ISBN-10: 1111990840

Hamilton MS (2012) Energy policy analysis: a conceptual framework. M.E. Sharpe. ISBN-13: 978-0765623829; ISBN-10: 076562382X

Hamilton County, Ohio (2001) Rules and regulations, the metropolitan sewer district of greater Cincinnati. http://www.msdgc.org/downloads/about_msd/legal_and_organizational_documents/msd-rules-regulations/rules_and_regulations.pdf. Accessed 30 Sept 2014

Hamilton County, Ohio (2009) Rules and regulations, hamilton county storm water district. http://www.hamilton-co.org/stormwater/HCSWD_Rules_And_Regulations.htm. Accessed 30 Sept 2014

Hansen AL (2010) The organic farming manual: a comprehensive guide to starting and running a certified organic farm. Storey Publishing, LLC. ISBN-13: 978-1603424790; ISBN-10: 1603424792

Heckel P, Dombek T (2009) Monitoring inorganic compounds. In Satinder Ahuja (ed) Handbook of water purity and quality. Elsevier, Science and Technology Books. Burlington

Heckel PF and LeMasters GK (2011) The use of AERMOD air pollution dispersion models to estimate residential ambient concentrations of elemental mercury. Water Air Soil Pollut 219:377–388.

Heckel PF, Keener TC, LeMasters GK (2013) Background soil mercury: an unrecognized source of blood mercury in infants? Online J Soil Sci 3:23–29. doi:10.4236/ojss.2013.31004

Lomborg B (2001) The skeptical environmentalist. Cambridge University Press. ISBN-10: 0521804477; ISBN-13: 978-0521804479

Muller RA et al (2013) Decadal variations in the global atmospheric land temperatures. J Geophys Res Atmos 118.11:5280–5286

National Abortion Federation (2014) About abortion. http://prochoice.org/education-and-advocacy/about-abortion/

Nice K, Bryant CW (2000) How catalytic converters work 08 November 2000. HowStuffWorks.com. http://auto.howstuffworks.com/catalytic-converter.htm. Accessed 30 Sept 2014

North Plains Groundwater Conservation District (2014) Hydrology & groundwater resources 2013–2014. www.northplainsgcd.org. Accessed 30 Sept 2014

Patel MR (2006) Wind and solar power systems: design, analysis, and operation, 2nd edn. CRC Taylor & Francis. ISBN-10: 0-8493-1570-0, ISBN-13: 978-0-8493-1570-1

Philander SG, Holton JR (2003) El Nino, La Nina, and the Southern oscillation index. Reprint edition. Grand Central Publishing. ISBN-10: 0446679925; ISBN-13: 978-0446679923

Schobert H (2013) Chemistry of fossil fuels and biofuels. Cambridge University Press. ISBN-13: 978-0521781268; ISBN-10: 0521804841

Seinfeld JH, Pandis SN (2006) Atmospheric chemistry and physics: from air pollution to climate change, 2nd edn. Wiley-Interscience. ISBN-13: 978-0471720188; ISBN-10: 0471720186

State of California (2012) Carbon Cap-and-Trade Program. http://www.arb.ca.gov/cc/capandtrade/capandtrade.htm. Accessed 30 Sept 2014

US Department of Energy (USDoE), Energy Information Administration (2013) Frequently asked questions. http://www.eia.gov/tools/faqs/faq.cfm?id=23&t=10. Accessed 30 Sept 2014

USEPA (2010) Renewable Fuel Standard Program (RFS2) regulatory impact analysis. EPA-420-R-10-006. http://www.epa.gov/otaq/renewablefuels/420r10006.pdf. Accessed 30 Sept 2014

USEPA (2012-12-4). Effects of acid rain. http://www.epa.gov/acidrain/effects/surface_water.html. Accessed 30 Sept 2014

USEPA (2014) Northern Gulf of Mexico Hypoxia http://water.epa.gov/type/watersheds/named/msbasin/zone.cfm. Accessed 30 Sept 2014

USEPA (2013) Proposed carbon pollution standard for new power plants. http://www2.epa.gov/carbon-pollution-standards/2013-proposed-carbon-pollution-standard-new-power-plants. Accessed 30 Sept 2014

US Government (USG) (1941) Flood Control Act of 1941 Public Law 228. http://www.fws.gov/habitatconservation/omnibus/r&ha1941.pdf. Accessed 30 Sept 2014

US Government (USG) (2007) Energy independence and security act of 2007 public law 110-140 (2007). http://www.gpo.gov/fdsys/pkg/BILLS-110hr6enr/pdf/BILLS-110hr6enr.pdf. Accessed 30 Sept 2014

US Department of the Interior (USDoI) (2008) Formal endangered species act consultation on the proposed coordinated operations of the Central Valley Project (CVP) and State Water Project (SWP), 81420-2008-F-1481-5. http://www.fws.gov/sfbaydelta/documents/swp-cvp_ops_bo_12-15_final_ocr.pdf. Accessed 30 Sept 2014

US Department of Justice (USDoJ) (2014) US Ninth Circuit Court of Appeals opinion determining a federal plan of protection for California's vital Delta ecosystem; Delta smelt biop ruling, Case #11-15871. http://docs.nrdc.org/water/wat_14031301.asp. Accessed 30 Sept 2014

Wilcox JR, Theodore L (1998) Environmental and engineering ethics. Wiley. ISBN 0-471-29236-2

Wright PH, Ashford NJ, Stammer RJ (1997) Transportation engineering: planning and design, 4th edn. Wiley. ISBN-13: 978-0471173960; ISBN-10: 0471173967

Chapter 2
Combustion

Abstract Combustion is the process that separates carbon atoms from fuel in the presence of oxygen. It is an exothermic reaction; energy is released as heat and light. The by-products of complete combustion of a pure fuel containing only atoms of carbon and hydrogen are carbon dioxide (CO_2) and water (H_2O). Biofuels are considered renewable fuels because they are made from carbon-containing substances that can be replenished annually. Biofuels are made from plant oils and sugars, the most common being ethanol and biodiesel. Fossil fuels are combustible substances created by geologic forces over time. Solid wastes containing carbon can also be burned.

Keywords Carnot cycle · Biodiesel · Kyoto protocol · Paper pellets · Co-generation · Ethanol

Combustion is the process that separates carbon atoms from fuel in the presence of oxygen. It is an exothermic reaction; energy is released as heat and light. The by-products of complete combustion of a pure fuel containing only atoms of carbon and hydrogen are carbon dioxide (CO_2) and water (H_2O) (Glassman et al. 2014; Baukal 2012; Green and Perry 2007). This chapter contains descriptions of combustible fuels, such as biofuels, fossil fuels and solid waste.

2.1 Biofuels

The Energy Independence and Security Act of 2007 provided millions of dollars to develop alternative fuels (USG 2007). Biofuels are considered renewable fuels because they are made from carbon-containing substances that can be replenished annually. Ethanol and biodiesel are liquid biofuels. Biomass means plant or animal material (Schobert 2013). Biomass burning refers to the deliberate use of combustion to reduce the volume of biological solid waste. For example, in Brazil, the plant material left over after harvesting sugar cane is incinerated to generate electricity. Biomass contains elements other than carbon and hydrogen; therefore, biomass

P. E. Heckel, *The Ethics of Energy Sustainability,* SpringerBriefs in Environmental Science, DOI 10.1007/978-94-017-9701-6_2

burning releases compounds other than water and carbon dioxide. Released carbon particulates are commonly called soot.

According to carbon commodities traders, plants grown and burned within 12 months are not counted in a country's carbon budget. This accounting trick encourages biomass burning to generate electricity regardless of the accompanying release of carbon particulates and other chemical compounds. Some incinerator facilities efficiently and completely combust biomass, emit water and carbon dioxide, and control the release of other compounds, including ash. Some incinerator ash contains hazardous compounds.

In countries lacking modern farm machinery, stubble left in the field after harvest is burned rather than tilled. Open burning is an incomplete combustion process that releases various gases and unburned carbon-containing soot particles. Think of forest fires. Soot is a pulmonary irritant which contributes to the incidence of asthma. Some types of soot are carcinogenic. Once in the atmosphere, they serve as cloud concentration nuclei (Chavez et al. 2013; Scholtens et al. 2014). In the upper atmosphere, carbon-based particulates may contribute to global cooling by absorption of solar radiation.

2.1.1 Biodiesel

Biodiesel is made from animal fats or vegetable oils through a process called trans-esterification. The fatty acids react with either ethanol or methanol. There are five common method of production. Byproducts include soap and glycerol with some excess alcohol, and trace amounts of water. Early reports of biodiesel production focused on individuals who collected waste fats from local restaurants and converted them.

2.1.2 Ethanol

Biofuels are made from plant oils and sugars, the most common being ethanol and biodiesel. To reduce transportation-related emissions of CO_2 in the United States, the USEPA mandated that ethanol be blended with gasoline. At the time, it seemed like a win-win situation. Global Warming activists cheered. Pacifists supported the ethanol mandate because it supposedly reduced the amount of oil imported from the Middle East. Planters celebrated because it created demand for corn. Academia rejoiced at the availability of federal dollars for renewable fuels research.

The chemical properties of ethanol are of concern. Ethanol cannot be shipped by pipeline because it is corrosive; it is trucked. Automobile engines may be damaged when the concentration of ethanol exceeds 10 % per gallon of gasoline. The mileage per gallon of blended fuel is less because the energy potential of ethanol is lower than gasoline. The BTU equivalent cost to make, transport, and blend one gallon of ethanol is 0.9 gallons of gas. However, 10 % of a million is 100,000. Blending ethanol with gasoline replaces a daily average of 36,851,000 gallons of gas, approximately 877,000 barrels of oil (USDoE 2013).

In the United States, ethanol is made primarily from corn, because of its high sugar content, although other plant material such as swtichgrass, sugar cane, and algae can be used (USEPA 2014a). Much of the corn in the Great Plains region, especially Iowa, is now grown for the ethanol market. Corn is also sold as livestock feed. As the price of corn increases, the price of milk and meat increases at the grocery store. Corn farmers point out that the brewers grains left over after the production of ethanol are sold as livestock feed. Nothing is wasted.

Ethanol production, whether for liquor or fuel, is water-intensive. It requires between 3 and 5 gallons of water for each gallon of ethanol produced. http://web.extension.illinois.edu/ethanol/wateruse.cfm. This creates an ethical dilemma for farmers in the Great Plains region and parts of Texas which depend on irrigation water from an underground aquifer which is not replenished when it rains. Naturally, ethanol plants are sited on several large rivers that cross the Great Plains. The Missouri River originates in Montana, crosses North and South Dakota, and forms the western border between Nebraska and Iowa and a portion of the border between Kansas and Missouri before it crosses Missouri to meet the Mississippi at St. Louis. The Mississippi River originates in Minnesota and forms the eastern borders of Iowa, Missouri, and Arkansas and bisects Louisiana. Water removed from the river is not returned as treated effluent, rather some evaporates and some remains in the distillers grains, which may be sold for livestock feed. Some downstream communities complain about the reduced volume available for their use.

The ethanol boom has made it financially attractive to cultivate parcels with marginal soils or challenging topography. Farmers who were barely scraping by have been able to replace broken equipment and pay off mortgages. Agribusinesses have made millions from producing corn for ethanol. These people support the EPA mandate.

Political Action Committees (PACs) are attempting to sway the debate. Conservationists in Prairie States have been dismayed to discover that abandoned agricultural land previously enrolled in the Conservation Reserve Program to preserve the native grasses of the prairie have been planted in corn (USEPA 2010). Bottom land that attracted migratory water fowl, because it was too wet to plow, has been filled and tilled. Organizations such as Ducks Unlimited and the Audubon Society have stepped up their fundraising efforts to preserve wetlands. Others want to reintroduce the buffalo to the plains. Land ownership is the key for conservation and many conservationists prefer public ownership of selected sites. Government entities meanwhile are assailed by increasing demands for services that have been financed by lease payments for growing, grazing or mineral rights. Which public good trumps all others?

The media has given much attention to the food for fuel issue. Has the ethanol mandate actually increased food prices? Looking at Table 2.1, the value of a bushel of corn in Oct 2013 is higher than it was in Oct 2009, but lower than a year ago. The price for other grains has also fluctuated. Is this a function of land use, crop yields, weather, or public policy? Prices are given in metric tonnes (metric

Table 2.1 Comparison of Grain Prices*

	2009	2010	2011	2012	2013
Rice	566.25	533.13	602.14	584.74	453.26
Soybeans	354.86	427.18	446.02	565.53	472.83
Wheat:	198.85	270.23	289.01	358.20	325.69
Sorghum	159.05	201.04	263.67	283.07	205.25
Corn	167.22	235.70	274.78	321.63	201.75
Barley	130.55	174.61	208.68	250.04	159.28

*price given in US dollars/metric ton as of Oct

ton = 1000 kg = 2025 lbs.), which is slightly larger than an English ton at 2000 lbs. The number of bushels per metric tonne depends on the size and weight of the harvested grain. The price reflects the weather more than any other factor (Index Mundi 2014).

2.1.3 *Wood*

Wood is a biofuel. It is considered a renewable resource; however, trees are harvested in 5-10-20 year cycles rather than annually. To account for the carbon emitted when burned, the amount of carbon sequestered during the current/last year can be subtracted from the total carbon content of the tree products burned. The benefits of trees include fuel, shade, oxygen formation, soil enhancement through root growth, some reduction of ambient particulate concentrations and wind erosion, absorption of storm water, and artistic appeal. Naturally, trees have other uses: building elements, paper, fabrics, cellulose, food, various chemical compounds, mulch, etc. The Biofuels Center of North Carolina advocates harvesting wood debris for fuel rather than using a prescribed burn, which may kill off endangered wildlife (Wall 2011)

The growth rate of trees/plants is positively correlated with the ambient concentration of CO_2. Tree rings provide an indication of the amount of biomass added and carbon sequestered each year. A comparison of growth rings shows that on average urban trees grow slower than rural trees, experience more stress, and are likely to die sooner.

Trees in an urban environment may negatively impact infrastructure. Roots cause upheaval/cracks of sidewalks and pavements and penetrate water, sewer, and septic lines. Falling branches/trees may damage power lines or property. Thus, an urban forestry management program requires regular inspection, pruning and replacement of tree stock. Many municipalities establish separate landfills for plant material.

In some areas, trees may enhance landscaping and may improve property values. For example, a healthy 50-year old oak located in the front yard of a suburban residence may be highly valued, but the same tree planted next to a structure or a driveway may be a nuisance. Farmers often use stands of trees along the western property line to prevent soil erosion from wind.

There are some advantages of planting trees in the median of an interstate highway. A stand of trees may improve safety by preventing errant drivers from accidentally driving into oncoming traffic, or by blocking the glare of headlights. The trees provide habitat for birds and other animals. Pollen, seeds, and insects are often carried on the wind; trees slow the wind and therefore, may slow the advancement of some non-native species. During a severe weather event, portions of the road may become blocked by downed trees. It is extremely rare for communities to either harvest fallen timbers or hay from the median strips. Perhaps that will change.

The most common ethical challenge when promoting biofuels is to present the complete picture. The cost of alternative fuels exceeds the cost of petroleum-based fuels in almost every instance. Different procedures are required to handle and store alternative fuels; they are not always onerous. The performance of these fuels in existing machines does not meet the same standards. Undoubtedly the technology will change. The primary benefit is that we are developing alternatives now rather than waiting until the supply of fossil fuels in threatened. Achieving energy independence is a noble goal for our domestic agenda.

2.2 Fossil Fuels

Over geologic time, plants and animals have lived and died. Natural forces have changed and shaped the crust of the earth. Continents have broken and drifted, creating mountain ranges and other land forms. The climate has changed from ice age to temperate in an unrelenting cycle (Lomborg 2001). Folding crust and massive ice sheets have created compressed carbon deposits which we now burn: coal, gas and oil.

2.2.1 Coal

Coal is thought to be the most abundant fossil fuel with reserves in the United States estimated sufficient for two hundred years or longer, a mere moment in geologic time. Communities of free men extract coal by hand in shafts dug deep into the earth. Their culture is full of stories and song about the miners and mine disasters, young widows and black lung disease. Strip mining near surface deposits seems in contrast to be safer and healthier. The anti-coal policies of the Obama Administration intentionally discourage demand in the United States for this abundant, cheap energy source; however, the Fukushima nuclear disaster in Japan has led to an increase in the demand for coal worldwide.

The discovery of coal as a fuel source is not recorded in any history book; however, the similarity between charred wood and coal is hard to miss. From camp fire to cook fire to hearth, coal burned hotter and longer than wood. Denser and lighter in weight, coal fueled the industrial revolution.

The chemical by products of coal combustion depend on the amount of available oxygen and the temperature of the fire. With insufficient air, deadly carbon monoxide (CO) and hydrogen gas (H_2) are produced. With excess air, carbon dioxide (CO_2) and water (H_2O). Other elements in coal include sulfur (S), and nitrogen (N) with traces of mercury ($Hg+$), chlorine ($Cl-$), arsenic (As), selenium (Se) and other elements (Glassman et al. 2014; Baukal 2012; Green and Perry 2007).

Under the CAA, areas that are not in attainment for the NAAQS must evaluate all proposed new sources or major modifications to existing sources to avoid further deterioration of the ambient air quality. In effect, the NAAQS limits the development of new coal-fired power generating facilities. Air pollution dispersion models such as AERMOD and CalPuff predict the impact of adding another emissions source to a specific location by estimating ambient concentrations of specific pollutants of interest. These models require access to geographic and land use data, chemical transport principles, and meteorological data.

Utilities are highly regulated to prevent price gouging. The cost of adding pollution control equipment such as scrubbers and electrostatic precipitators and carbon capture and storage (CCS) technology to proposed coal burning facilities is likely to exceed the expected profit margin of the operation. The utilities may then apply for a rate increase to make their stock attractive to shareholders. The 2013 version of USEPA proposed regulations limits every emission except water emissions from coal-fired power plants (USEPA 2013).

Once the President announced a war on coal, savvy investors sold coal stocks. Coal companies found it harder to access capital. Jobs in coal mining towns dried up. The ethical question is whether this energy policy was intended to move the country closer to energy independence or not. Was it part of a long term Green Energy strategy or simply market manipulation and political retribution?

2.2.2 *Natural Gas*

A molecule of natural gas contains carbon and hydrogen atoms. When completely burned in the presence of oxygen, carbon dioxide (CO_2) and water (H_2O) are produced. The most common natural gas is methane (CH_4), consisting of one carbon atom and four hydrogen atoms. CH_4 is formed during the decomposition or digestion of plant and animal materials. CH_4 is a highly reactive greenhouse gas, estimated to be 50 times more harmful in the stratosphere than CO_2. Atmospheric collisions between CH_4 and the hydroxyl radical (OH) produce water vapor. Natural gas is odorless and colorless. For safety, an odorous chemical compound hydrogen sulfide (H_2S) is added to natural gas carried through distribution systems. Domestic uses include home heating systems, gas stoves for cooking, gas fireplace inserts, and gas water heaters. CH_4 is also used as a transportation fuel. Since it contains fewer carbon atoms than gasoline, cars using a blend of ethanol and gasoline get fewer miles per gallon than cars burning regular gasoline. Removing H_2O from landfill gas is a technical problem generally solved by condensing out the H_2O. Ensuring that the CH_4 meets contract specifications is a basic ethical decision.

2.2.3 Petroleum

Petroleum is a fossil fuel formed by geologic events acting on plants, such as algae, and animals, such as zooplankton, trapped under ancient sedimentary rock deposits. The process of recovering the deposits is called "drilling for oil". Petroleum products include liquids and gases containing carbon compounds. Within the geologic formation, petroleum can be dispersed in sand, trapped in shale, or pooled. Early recovery focused on liquid oil. Current methods such as fracking can extract oil and gas from sand and shale.

During the refining process, gases containing different numbers of carbon atoms are separated by their boiling point and captured. The name of the gas is related to the number of carbon atoms and/or its molecular structure. For instance, octane contains eight carbon atoms. The lighter compounds are used for transportation fuels. Heavier liquid compounds are used for lubrication of machinery. Paraffin has common household uses, for instance in candles or to seal jars of jam. The heaviest compounds are called bitumen, asphalt, pitch or tar and are used in road construction or as a sealant. The boiling point of these heavy compounds is above 500° F. During the combustion process, the carbon atoms are released from the fuel and react with oxygen atoms. This can be a multi-step process.

Petroleum-based fuels are ideal for transportation because they are safe and they have a high heat capacity per ton, measured in BTUs. This means that more fuel can be carried in a smaller space. The improvement in air quality in Pittsburgh, PA is partially attributed to the switch from coal-fired boilers to gas turbine engines on river barges and partially attributed to the economic ruin of the steel industry in the United States. Train safety was likewise improved when diesel-powered engines replaced coal and wood-fired boilers.

Petroleum-based fuels are used in aviation. Diesel fuel is adequate for some compression ignition engines, while others respond better to kerosene or jet fuel with a lower viscosity. Piston-engines use a high-octane gasoline. Environmental concerns with air travel include noise during take off and landing and when jets flying faster than MACH 1 break the sound barrier; the emission of air pollutants, in particular unburned hydrocarbons, NOx and SOx; and soil and water pollution from de-icing compounds, oil and fuel.

2.3 Solid Waste

Solid waste includes human excrement, food waste, industrial wastes, yard wastes, assorted garbage, combustion by-products, and all sorts of discarded objects, such as clothing, furniture, electronics and appliances, tires, and construction debris. Although some countries encourage residents to recycle their cast-off items and to compost their food and yard wastes, a large volume of waste remains. Incineration is a method used to burn solid wastes. It produces high temperatures. This heat may be used to generate steam.

Medical incinerators are used to burn human tissue, plastics, sponges, blood-contaminated substances, radioactive substances, and other biohazards. Emissions are regulated to limit the release of radioactive particles as well as the criteria pollutants. In 2014, the National Health Service (NHS) in Great Britain was roundly criticized for burning aborted fetuses and other human tissues in a waste-to energy incinerator without informing patients of other disposal options, including burial. http://www.telegraph.co.uk/health/healthnews/10717566/Aborted-babies-incinerated-to-heat-UK-hospitals.html

Solid waste incinerators are used to burn garbage. The first municipal solid waste incinerators created tons of ash containing hazardous metallic compounds. This problem was alleviated in part by removing steel, aluminum, batteries, and electronic waste from the garbage.

Some waste water treatment centers incinerate sludge. The City of Palo Alto CA sells their ash to a mining company which extracts silver and gold from Silicon Valley sludge (Bishop 1995). Each of these incinerators emits carbon dioxide and water vapor while reducing the volume of solid waste that is transported to landfills.

2.3.1 Tire-Derived Fuel

In the United States, approximately 290,000,000 tires are scrapped annually (USEPA 2014b). Discarded tires from motorized vehicles are a persistent environmental and health hazard. Discarded tires clutter the landscape and occupy valuable landfill space. Water accumulates inside tires and provides a breeding ground for mosquitoes. How can tires be eliminated as waste?

Although the carbon content of rubber made it a potential fuel, research into tire disposal and reuse identified two problems: the steel belt and sulfur. A machine was developed to remove the steel belt from the tire. The tire was then placed on a conveyor to a second machine which chopped it into small (1") pieces. These pieces were sold as a landscape fill material for playgrounds. To make this process economically attractive, government entities collected a tire disposal fee, which was used to offset tire collection costs.

The next phase of research focused on the use of these tire pieces as a fuel, either alone or combined with another solid fuel. Current NAAQS limit emissions of sulfur oxides. In some areas, burning tires alone emits more SOx than allowed by permit. The alternative is to either mix the tire chips with a low sulfur fuel and/or to install costly pollution controls, such as flue gas desulphurization equipment. Continuous emissions monitoring provides a record of emissions in real time. Plant operators can adjust the conveyor speed and fuel mix to minimize exceedances.

2.4 Additional Topics for Research

Resource Conservation and Recovery Act (RCRA), 40 CFR Part 264 Subpart O-Incinerators, 40 CFR Part 265 Subpart O-Incinerators, PM10, PM2.5, Haze, Landfill, Recycling, Carcinogen, Pulmonary diseases, Regional Haze Program, RIO Plus 20, Road ecology, Aquaponics, The Dune Trilogy, Appalachia, CO Monitors, Gypsum, Sorbents, Electrostatic precipitators, Dust cyclones, Bag house, Thermal oxidizers, Particulate scrubbers, Catalytic reduction, Vapor recovery systems, Abatement, Flue-gas desulphurization, Continuous emissions monitoring systems, Carbon capture, Coal gasification, "Clean coal" technology, Octane, Greenhouse gas, Marcellus shale, Barnett shale, Mitchell Energy, Fracking, Landfill gas, Coal gasification, Flares, Refinery, Keystone Pipeline, Trans-Alaskan Pipeline, North Slope, Valdez, Sulfur, SOx, NOx, Peak oil, British Thermal Unit (BTU), Igor Sikorsky, Hoppers, Avgas, Diesel fuel Toxic Substances Control Act, National Emissions Standards for Hazardous Air Pollutants, (NESHAP), Leachate Vulcanization, Rubber, Permit to emit, Title V, West Nile Virus, Malaria, Yellow fever, Disease vectors.

2.5 Assignments

1. What is the impact of the ethanol mandate on land use? The states producing corn for fuel are not able to use their land to produce other crops. Does this lead to a food shortage in those states? Explain.
2. As grain is shipped out, some foods are imported. Most farmers have a family vegetable garden. People in urban areas generally buy their food from grocers. Are fresh foods less available or shipped further because of the ethanol mandate?
3. Has the ethanol mandate affected the price of cane sugar? Beets? What is a sugar cartel?
4. If more fats and oils are diverted to produce biodiesel, the supply of what other products may be affected?
5. Ladybird Johnson, the wife of US President Lyndon Johnson, urged the passage of the 1965 Highway Beautification Act. How has the Act encouraged the cultivation and preservation of native plant species in your neighborhood?
6. United Airlines allows passengers to purchase credits to offset the CO_2 emissions attributed to their air travel. Who gets this money and how is it used?
7. Why are plant materials and yard wastes banned from sanitary landfills? Name three revenue-producing uses for these solid wastes.
8. Does your community collect and compost plant materials and yard wastes? What happens to the compost?
9. Why do municipalities value trees? Provide six reasons for planting trees in urban areas.

10. What is Methane clathrate? Where does it exist on earth? What problems are associated with its use as a fuel?
11. Summarize the myth and the reality that the Zionists made the desert bloom in Israel. How and where could those lessons learned be applied today?
12. What is the purpose of the organization "ILoveMountains.org"? How many of the supporters actually live in Appalachia?
13. What is black lung disease? Is there a cure?
14. Which is safer for miners, strip mining or shaft mining?
15. Describe three technologies to use coal for energy.
16. Research the profitability of utility companies that provide electricity in your state. The price they charge for electricity is limited by the Public Utilities Commission. How much is the dividend per share of stock?
17. The USEPA on September 20, 2013 proposed regulations to limit CO_2 emissions from power plants by requiring carbon capture and storage (CCS). The intention is to force operators of older coal-fired plants to retire those facilities, reducing electrical output by approximately 40% with an accompanying increase in utility rates. What is the state of the science of CCS? What geologic conditions are required to implement it?
18. The typical electric utility bill contains a charge for the electricity used plus taxes and fees. In some regions, the price of energy is higher during peak hours. Examine a local electric bill. How much of the bill covers the cost of the electricity? What is the purpose of the taxes and fees?
19. What is lignite and how is it different from bituminous coal? Where are different types of coal found? Compare the heating value in BTUs of coal to other fuel sources (biofuels and fossil fuels).
20. The USEPA has proposed a national regulation that requires carbon capture and storage for coal-fired plants. Utility XYZ is not affected by the regulation because it operates natural gas-fired plants, municipal waste incinerators, and a windmill farm. Utility provider DEF opposes the regulation. Utility DEF is operating several coal-fired plants, a hydroelectric facility, and a nuclear reactor. Utility XYZ serves a densely populated region that is flat, windy and has large reserves of natural gas. Utility DEF serves a sparsely populated mountainous region with ample coal reserves and several swift rivers. The citizens in the region served by Utility XYZ voted for the current administration. The citizens in the region served by Utility DEF voted for the opposing candidate. Discuss the ethics of the proposed regulation.
21. CASE STUDY A rural mountainous community in West Virginia owns a parcel of land with a large deposit of coal. A company in Germany wants to buy the coal for export to Germany because the Russians are threatening to disrupt their supply of oil. The deposit is located near the surface, making strip mining the most economical method to harvest the coal. The company promises to comply with all environmental restrictions. Once the coal is removed, they will develop the parcel for civic and commercial activity. The citizens support the proposal. An outside group opposes it; they want the land to remain undeveloped, but have not offered to buy it. On the worksheet, Fig. 2.1 list the pros and cons of this proposal. Should the town risk a lawsuit from the outside group?

West Virginia Coal

Health & Safety	Environment	Economy
Pro:	Pro:	Pro:
Con:	Con:	Con:

Fig. 2.1 West virginia coal

22. The Rumpke Sanitary Landfill in Ohio near Cincinnati has installed a methane recovery system to capture, clean and sell landfill gas. Research and summarize the operation. How is water removed from the gas? What is the lowest concentration of methane that can be economically recovered?
23. Methane gas is released from coal deposits. What safety precautions are available to protect miners from suffocation due to methane exposure? What safety precautions are used to prevent explosions in underground mines?
24. What is anaerobic digestion? How is methane gas recovered from manure? What type of animal operations can profitably recover methane?
25. How does your local waste water treatment facility handle sludge?
26. Fracking allows the removal of hydrocarbons from deposits which are too far underground to mine. Explain the fracking process. What is a ground water table?
27. A study of silicate exposure during fracking was conducted by the National Institute of Occupational Safety and Health and Schlumberger. The research team found that the exposure of workers to silicates was below existing threshold limit values (TLV). What is the TLV for silicate exposure? How is a TLV established? When during the fracking process are workers likely to be exposed to silicates?
28. CASE STUDY A widow has inherited the family farm, but very little money to care for it. The farm has 100 acres, 10 acres on each side, and has a spring that runs across the SE corner. Shale gas has been discovered on the north and west portions. An drilling company has offered her enough money to pay her taxes for ten years in exchange for the drilling rights. She would also receive royalties on the volume of gas extracted. There is an unoccupied farm house on the property near the spring. The company vows to meet all environmental regulations associated with fracking and to grade the property for agricultural

Fracking on the Family Farm

Health & Safety	Environment	Economy
Pro:	Pro:	Pro:
Con:	Con:	Con:

Fig. 2.2 Fracking on the family farm

use after the gas has been recovered. The field extends north and west and some of her neighbors have made agreements with this same company. Use worksheet Fig. 2.2 to identify pros and cons. Should the widow accept the offer?

29. Summarize the process of refining oil. Name the gases and other products that are separated from the oil in the distillation column. What is the boiling point of each? Why is the boiling used? http://science.howstuffworks.com/environmental/energy/oil-refining2.htm
30. What is the difference between a turbine and a generator?
31. How is jet fuel different from diesel?
32. What are the advantages of using warm mix asphalt for roads (as compared to hot mix asphalt)?
33. What is the Alaskan Native Claims Settlement Act?
34. The Trans-Alaskan Pipeline runs from Prudhoe Bay to Valdez. Why was Valdez chosen as the southern terminal? How many miles of the pipeline cross permafrost? What was the issue with the caribou and how was it resolved? Where are the natural gas deposits in relation to the current pipeline? Show these areas on a map of Alaska. Who has jurisdiction over the Trans-Alaskan Pipeline and why?
35. Summarize the BP oil spill incident in the Gulf of Mexico. Who got the money? Who went to jail? Why?
36. Describe the safety protocol for entering a confined space. List three types of "sniffer" equipment.
37. Five or more environmental impact studies were conducted on the proposed Keystone Pipeline. Who conducted the studies and what were their conclusions? What other options exist for refining the oil from Canadian oil sands?
38. What is the difference between "sweet" and "sour" oil?

Airport Expansion

Health & Safety	Environment	Economy
Pro:	Pro:	Pro:
Con:	Con:	Con:

Fig. 2.3 Airport expansion

39. CASE STUDY A town of 100,000 residents is wooing a Fortune 500 Company. The city expects that 5000 good-paying jobs will be created. Included in the package of incentives is a commitment to build a runway at the regional airport capable of handling the Boeing 787 Dreamliner. The airport owns a multi-acre buffer around the existing facilities and no additional land is needed for the expansion. Outside of the property on one side is dry stream bed which captures runoff whenever the area receives more than 1" of rain during a storm. The town water supply is not expected to be affected by the airport expansion. The air quality in town is very good, with no attainment issues. Is this a sustainable situation? Assign this case study to two teams, one representing the company and the other representing the town. List the pros and cons on worksheet Fig. 2.3. What other assumptions are needed?
40. What is Acid Rain? How is acid rain related to fish kills? Where is acid rain a problem?
41. What was the impact of vulcanization on the automotive industry?
42. How are discarded tires and West Nile Virus related?
43. What is the fee to dispose tires in your area? How is that money used?
44. A tire recycling facility is planned for a city of 1 million people. How many tires can the facility expect to recycle each year? The state collects a \$ 3/tire disposal fee. The recycling facility receives \$ 2.50/tire. Chips from the tires are sold for \$ 10/metric t. Steel belts from the tires are sold as scrap for \$ 0.25 each. What is the expected revenue for this facility? If the foreman makes twice minimum wage and the other workers make minimum wage, how many full time employees can the facility afford to hire? Be sure to include the employer contribution for wage and earnings taxes, 8 paid holidays per year, and \$ 12,000 in health insurance per employee.

45. A municipality plans to cover their 6 lane running track in pressed, recycled tires. The cost of the tire chips is \$ 10/metric tonne. A six inch deep cover is recommended. Two men from Beautify It, a landscaping company, can spread the chips at a rate of a truck load (2 metric tonne) every 2 hour. The men make minimum wage. Beautify It pays the usual employer contribution for wages and earnings taxes, 8 paid holidays per year, and \$ 12,000 in health insurance per employee. They also pay a bonding fee of \$ 0.10/\$ 1000 bid and carry a \$ 1,000,000 in liability insurance. The cost to seal the surface is \$ 10/sq ft. They estimate a 10 % profit. What is their bid?
46. Limestone is a buffering agent. How much calcium carbonate is required to change the pH of a cubic meter of water from 6 to 7?

References

Baukal CE Jr (2012) The John Zink combustion handbook (Industrial Combustion). CRC Press

Bishop P (1995) Municipal sewage sludge: management, processing and disposal, vol 4. CRC Press. ISBN 0877629307, 9780877629306

Chavez WL et al (2013) Identification of asthma-specific cis-regulatory DNA by epigenomic profiling. Am J Respir Crit Care Med 187:A1211

Glassman I, Yetter R, Glumac N (2014) Combustion, 5th edn. Academic. ISBN-10: 012407913X; ISBN-13: 978-0124079137

Green D, Perry R (2007) Perry's chemical engineers' handbook, 8th edn. McGraw-Hill Professional. ISBN-10: 0071422943, ISBN-13: 978-0071422949

Index Mundi (2014) Commodity price indices. http://www.indexmundi.com/commodities/?commodity=rice&months=60

Lomborg B (2001) The skeptical environmentalist. Cambridge University Press. ISBN-10:0521804477; ISBN-13: 978-0521804479

Schobert H (2013) Chemistry of fossil fuels and biofuels. Cambridge University Press. ISBN-13: 978-0521781268; ISBN-10: 0521804841

Scholtens S et al (2014) Novel childhood asthma genes interact with in utero and early-life tobacco smoke exposure. J Allergy Clin Immunol 133.3:885

US Department of Energy (USDoE), Energy Information Administration (2013) Frequently asked questions. http://www.eia.gov/tools/faqs/faq.cfm?id=23&t=10. Accessed 30 Sept 2014

USEPA (2010) Renewable Fuel Standard Program (RFS2) regulatory impact analysis. EPA-420-R-10-006. http://www.epa.gov/otaq/renewablefuels/420r10006.pdf. Accessed 30 Sept

USEPA (2013) Proposed carbon pollution standard for new power plants. http://www2.epa.gov/carbon-pollution-standards/2013-proposed-carbon-pollution-standard-new-power-plants. Accessed 30 Sept 2014

USEPA (2014a) Renewable and alternative fuels. http://www.epa.gov/otaq/fuels/alternative-renewablefuels/index.htm

USEPA (2014b) Scrap tires. http://www.epa.gov/osw/conserve/materials/tires/basic.htm

US Government (USG) (2007) Energy Independence and Security Act of 2007 Public Law 110–140. http://www.gpo.gov/fdsys/pkg/BILLS-110hr6enr/pdf/BILLS-110hr6enr.pdf. Accessed 30 Sept 2014

Wall S (2011) The biomass endeavor. Wildlife in North Carolina. Feb. pp 14–19 downloaded from. http://www.biofuelscenter.org/north-carolinas-biofuels-campus. Accessed 28 Sept 2014

Chapter 3
Heat Exchangers

Abstract Heat exchangers are used to transfer heat from a hot liquid or gas source to a cooler liquid or gas sink. This principle is applied to boil water and to freeze ice, to moderate the temperature in our homes and to desalinate water (Green and Perry 2007). Under certain conditions, energy from the sun can be used instead of combustion to accomplish the same objective.

Keywords Refrigerant · Desalination · Hot springs · Thermal energy · Oil coolers · Solar heaters

Heat exchangers are used to transfer heat from a hot liquid or gas source to a cooler liquid or gas sink (Cengel and Ghajar 2014). The two fluids remain separated by the walls of their containers. For example, natural gas is burned in a small chamber, heating the walls of the chamber. A large box outside the chamber contains air that is cooler than the air inside the chamber. A fan is used to move the air across the walls of the small chamber, thereby transferring heat from the small chamber to the air in the box. The air in the chamber and the air in the box do not mix. This forced-air process is used to heat homes.

A solar hot water heater operates in much the same way (Cengel and Ghajar 2014). A small pipe connected to an external source of cold water runs through a larger diameter cylinder full of air. This assembly is placed in the sun. The sun heats the air and the air heats the water. A pump or gravity is used to move the water into an insulated storage device. The system is regulated so that the pump turns off before the storage device overflows. The air and the water do not mix.

If two liquids are used, a tube of hot liquid flows in one direction. It touches a second tube containing a cold liquid which flows in the opposite direction. Heat is transferred across the intersection of the two tubes. In the most efficient systems, the surfaces between the two fluids are tightly fused so that they act as a common wall. The end view of the tube assembly resembles a figure eight. Again, there is no mixing of fluids.

P. E. Heckel, *The Ethics of Energy Sustainability*, SpringerBriefs in Environmental Science, DOI 10.1007/978-94-017-9701-6_3

3.1 Ocean Thermal Energy Conversion (OTEC) Systems

Many coastal areas lie adjacent to very deep oceans (Garrison 2012). The temperature difference between the hot surface water and the deep ocean water can be 20 degrees (F) or more in tropical climes. This temperature difference is sufficient to drive a heat engine using heat exchangers (Cengel and Boles 2014). This technology was proven in 1974 at an OTEC facility located off Kona, Hawaii and administered by the Natural Energy Laboratory of Hawaii Authority (NELAH 2014) Current OTEC facilities are generally used to desalinate water since the supply of fresh can be limited in these areas of the globe. The success of the OTEC concept indirectly led to the development of the vertical well version of the geothermal heat pump solution to reduce home heating and cooling energy demand.

3.2 Low Grade Geothermal Systems (LGG)

A low grade geothermal heat pump system (GHPS) is also called an Earth-Coupled Geothermal System or a Ground Source Heat Pump. This system takes advantage of the fact that in every solid land location at some distance below the surface of the Earth, the land never freezes. In the continental United States, this depth is approximately six feet below the surface.

There are basically two configurations of GHPS, vertical or horizontal. The horizontal field is laid in trenches below the frost line. The trench is backfilled with the original material. At the Glen Helen Nature Center in Yellowsprings, OH, the horizontal field was laid below the gravel parking lot. In areas where a horizontal field is not practical, a vertical shaft is dug. The pipes are replaced with a coil. Many of these systems deliberately penetrate the ground water table and consider the temperature of the ground water as a design parameter. The length of the pipe/coil is determined by calculating the heat loss (Cengel and Ghajar 2014). A heat exchanger is used to add/extract heat between the inside building temperature and the temperature of the liquid inside the pipe. It is significantly cheaper to heat a structure to a comfortable temperature when the temperature difference is within 20 degrees (F) of the starting temperature.

3.3 Geothermal Heat

Some water lies beneath the surface of the Earth. In places where the tectonic plates overlap, water heated by the molten core is often accessible. Early peoples attributed healing and supernatural events to those locations. Well-known hot springs include Beppu, Japan; Jigokudani Monkey Park, Japan; Rincón de la Vieja, Costa Rica; Valley of Geysers, Russia; El Tatio, Chile; Rotorua, New Zealand; Huanglong, China; Pamukkale, Turkey; Dallol, Ethiopia; Blue Lagoon, Iceland; and Yellowstone,

United States. Latitude appears to not be a factor; hot springs are found in Banff, Alberta, Canada and significantly further south in Warm Springs, Arkansas, United States. Uganda in Africa has many. Sometimes the water erupts periodically, such as the geyser Old Faithful at Yellowstone National Park in the United States and the geysers of Haukadalur, Iceland. These natural formations are found on every continent.

The commercial advantage of using this type of geothermal heat depends on the temperature of the water and the local climate. In some areas, the hot water is used in a spa-like setting, appealing to religious pilgrims and the tourist trade, such as in Aquae Sulis (Bath), Britain. The oldest hot tub is located on Mount Li in China. Iceland and Turkey are particularly blessed with hot springs which are used to heat entire villages. There are industrial applications in New Zealand.

3.4 Additional Topics for Research

Refrigerant, Montreal Protocol on Substances that Deplete the Ozone Layer, Vienna Convention for the Protection of the Ozone Layer, Chlorofluorocarbons (CFCs), Hydrochlorofluorocarbons (HCFCs), Society of Automotive Engineers (SAE), Iceland Deep Drilling Project, District heating, Staufen im Breisgau, Magma, Volcanoes

3.5 Assignments

1. What type of refrigerant is most commonly used in geothermal heat pumps and why? How is a refrigerant's "phase change" used in a heat exchanger?
2. What is the difference between Chlorofluorocarbons (CFCs) and Hydrochlorofluorocarbons (HCFCs)?
3. Why is tropospheric ozone bad and stratospheric ozone good? How does ozone move between atmospheric layers?
4. Identify three commercial applications for heat exchangers. Which fluids are used?
5. Using a topological map, identify locations which are likely candidates for OTEC technology. Could the concept work in reverse using warmer water below the ice-clogged surface above the Artic Circle? Why or why not?
6. How is OTEC used to desalinate water?
7. What did Robert C. Webber demonstrate in his Indianapolis, IN home in 1945? Visit http://welldrillingschool.com/courses/pdf/geothermal.pdf and take the test at the end of the paper.
8. Before refrigerators, people stored food in the ground after the harvest for later consumption. What is a root cellar and what types of foods can be stored? What is a spring house and what types of foods can be stored?
9. Explain the carnot cycle. What are the limitations of this theoretical cycle?

10. Toronto, CA has an historic section that attracts thousands of tourists each year. The city owns a number of the buildings. In an effort to reduce the cost to heat the buildings, all the single-paned windows were replaced with energy-efficient windows. This reduced the bill somewhat, but City council is considering the potential of LGG to save more money. Compare the vertical well vs. horizontal field LGG systems for this application. Which would you recommend and why?
11. A developer is building a new Platinum LEED subdivision on a 100 acre parcel, which was formerly farm land. Typical precipitation for this property is 39" per square foot per year. He plans to dig ponds to capture storm water runoff from the roads and grey water from the homes. The government is offering an attractive tax incentive to install LGG systems. He estimates there will be 4 homes per acre, with each home having a 2000 square foot footprint. Compare the vertical well vs. horizontal field LGG systems for this application. Which would you recommend and why? Prepare a site plan showing the LGG systems.
12. Which industries in New Zealand use geothermal heat?
13. Which cities in your country, if any, use geothermal heat for district heating?
14. What were the engineering challenges addressed by the Iceland Deep Drilling Project?

References

Cengel Y, Boles M (2014) Thermodynamics: an engineering approach, 8th edn., McGraw-Hill ISBN-10: 0073398179. ISBN-13: 978–0073398174

Cengel Y, Ghajar A (2014) Heat and mass transfer: fundamentals and applications McGraw-Hill, 5th edn. ISBN-10: 0073398187. ISBN-13: 978–0073398181

Garrison TS (2012) Oceanography: an invitation to marine science, 8th edn. Cengage Learning. ISBN-13: 978–1111990848; ISBN-10: 1111990840

Green D, Perry R (2007) Perry's chemical engineers' Handbook, 8th edn, McGraw-Hill Professional. ISBN-10: 0071422943, ISBN-13: 978–0071422949

Natural Energy Laboratory of Hawaii Authority (NELAH) (2014) http://nelha.hawaii.gov/. Accessed 30 Sept 2014

Chapter 4
Nuclear

Abstract Two processes, fusion and fission, are encompassed by the term nuclear energy. Both involve the manipulation of atoms. The association of nuclear energy with the bombs dropped by the United States over Japan during World War II has hindered the acceptance of this technology for peaceful purposes. Nuclear energy remains one of the safest methods of generating power due to redundant safety systems and technological improvements. Beneficial applications in medicine and space travel are often ignored by those who oppose the hazardous process of uranium enrichment.

Keywords Hunley · Voyager · Obninsk · Three Mile Island · Chernobyl · Fukashima

Nuclear power refers to energy released during either fusion or fission of atoms. Peaceful applications of nuclear energy are generally limited to heat and electrical power generation. The first nuclear reactor in the United States was built to produce weapons grade plutonium-239 for bombs (Rhodes 1986). Today energy from radioactive substances is used in submarines, power plants, medical devices and space ships.

In 1620 Cornelius Drebble built the first known prototype of a submersible vehicle. It required oars for propulsion. A century later, in 1747, Nathaniel Symons used leather bags filled with water for ballast and stability. David Bushnell demonstrated a screw-powered propulsion system on the Turtle in 1775. Another century passed. Problems remained. During the Civil War, the Confederate submarine the H. L. Hunley sank a Union warship in 1864. Other submarines built during that decade were powered by either compressed air or combustion. Twenty years later, the steam engine and batteries extended its range and usefulness for combat. The development of nuclear power in the 1950s accompanied by the mastery of chemical processes to extract oxygen from sea water launched the era of the modern submarine, which can remain submerged for months (Harris 2014).

After World War II, research into peaceful applications of nuclear technology continued at Oak Ridge National Laboratory. Electricity from a nuclear power plant

P. E. Heckel, *The Ethics of Energy Sustainability,* SpringerBriefs in Environmental Science, DOI 10.1007/978-94-017-9701-6_4

was first produced in 1948. The race was on. An experimental power generating station was built in Arco, Idaho, United States in 1951. This was followed in 1954 by a full scale operation in Obninsk, Soviet Union of Socialist Republics. Space travel was made possible by Plutonium-powered engines.

Radiation is used in medicine to either treat diseases or to obtain images of the inside of the body (Mettler Jr and Guiberteau 2012). The most common tool is the X-ray. Applications of X-ray technology include mammograms and images of the bones. Radiation kills cells, so targeted radiation therapy can shrink tumors. A radionuclide combined with other chemicals can be injected through the vein. Imaging devices reveal the path of the solution through the body. This procedure identifies arterial blockages and internal hemorrhages.

Even considering the incidents at Three Mile Island, Chernobyl, and Fukashima, nuclear energy remains one of the safest methods of generating power due to redundant safety systems and technological improvements. Due to the abundance of uranium feed stocks, the Carter Administration implemented policy to store used fuel rods rather than to reprocess them. This led to the closing of the Hanford and Fernald facilities and to extensive environmental clean up of both of those sites. After the Chernobyl incident, several research universities in the United States shut down their nuclear energy programs. The second Bush Administration supported commercial reprocessing of spent fuel rods to extract plutonium and other fissionable materials and initiated a process to license six new nuclear power generating plants under the Energy Policy Act of 2005 (GPO 2005).

After the Fukushima incident in 2011, the Obama Administration called for additional research into fuel rod reprocessing and carbon capture and storage (CCS) technology. The Japanese closed most of their nuclear power generation stations and increased their consumption of coal. The German government announced that it intended to close several of its facilities and approved licenses for next generation coal-fired power plants to meet future energy demand. In England, United Kingdom, public opinion shifted in favor of fracking for recovery of natural gas. The 2014 Soviet invasion of Crimea and subsequent 30-year agreement to sell its natural gas to China raised additional concerns regarding the global energy supply.

4.1 Additional Topics for Research

Yucca Mountain, Rare earth metals, Actinide, Plutonium

4.2 Assignments

1. What is the difference between alpha, beta and gamma radiation? Which is most harmful to humans?

Generation IV Nuclear Facility

Health & Safety	Environment	Economy
Pro:	Pro:	Pro:
Con:	Con:	Con:

Fig. 4.1 Generation IV nuclear facility

2. In the United States, President Jimmy Carter's nuclear energy policy was to not reprocess spent fuel rods because uranium was cheap and there were concerns about nuclear proliferation. What is considered "spent fuel"? What happened to this radioactive material? What were the motivating factors for the reconsideration of this policy by President George Walker Bush?
3. What are the commercial applications for MOx nuclear fuel?
4. How is a thermal reactor different from a breeder reactor?
5. What is the difference between a fusion and a fission reactor? What are the fuel requirements?
6. CASE STUDY Is nuclear energy sustainable? Use worksheet Fig. 4.1 to list the pros and cons of replacing an existing coal-fired power generating station with a Generation IV nuclear power generation station.

References

Government Printing Office (GPO) (2005) Energy Policy Act of 2005. http://www.gpo.gov/fdsys/pkg/PLAW-109publ58/html/PLAW-109publ58.htm

Harris B (2014) Submarine history. http://www.submarine-history.com/

Mettler FA Jr, Guiberteau MJ (2012). Essentials of nuclear medicine imaging, 6th edn. Saunders. ISBN-10: 1455701041. ISBN-13: 978–1455701049

Rhodes R (1986) The making of the atomic bomb. New York: Simon & Schuster. ISBN 0–671-65719–4

Chapter 5
Solar

Abstract Solar power includes passive systems that capture light or extract heat from solar radiation and active systems that convert solar radiation into electricity. Passive solar has been used for centuries. Active solar is a more recent development, which may rely on the use of rare earth elements to convert light into electricity.

Keywords Solar incidence · Heat island effect · Concentrated solar · Heliostat · Photochemical · Photovoltaic

5.1 Passive Solar

The term passive solar implies the direct use of solar radiation and light. Most commonly, energy from the sun is used to heat surfaces or substances or to provide light. Ancient architects incorporated passive solar light features into dwellings. Sacred spaces often included openings in structures to allow sunlight to strike an altar on a specific day of the year. For example, the Fort Ancient culture built the Great Serpent Mound in Ohio circa 1070 CE and held special ceremonies to mark the solstices (Carmean 2009). Stonehenge in Wiltshire, England, may have been constructed as a celestial observatory as early as 3000 BC.

Some architects consider latitude to reduce energy demand when designing homes. The latitude of a property is closely correlated with a known range of solar angles of incidence. In the Northern Hemisphere, the strongest sunlight falls on the southern face of a building at high noon on June 21. During daylight hours in the winter, a room on the south-side would be warmer than a room on the north-side. Skylights, mirrors and windows can be used effectively to reduce the amount of artificial lighting needed in interior spaces.

Light waves are reflected from dense materials such as rock, tile, brick, and concrete. These materials heat and cool slowly. Dark colors such as black, brown, and navy absorb radiant heat. Light-colored materials in shades of white reflect light rays (Green and Perry 2007). Close to the equator, houses may be constructed with two light stone walls separated by a sizable air gap to slow the transfer of heat from outside to inside spaces. Some houses contain a private, shaded courtyard.

P. E. Heckel, *The Ethics of Energy Sustainability,* SpringerBriefs in Environmental Science, DOI 10.1007/978-94-017-9701-6_5

Landscaping features, including plants, water, and surfaces can be used effectively to reduce energy demand. For example, deciduous trees provide shade in the summer but do not block the weaker winter sunlight. A row of trees can slow the wind and reduce soil erosion, with evergreen trees acting as a year-round natural fence. Dark mulch absorbs water and heat. White sand reflects light and allows water to drain. A concrete driveway reflects heat, and is suitable for tropical areas. An asphalt driveway absorbs heat, and is recommended for cold climates (Green and Perry 2007). Both are impervious surfaces. Water heats more slowly than soil. The temperature difference between a pond and the surrounding land creates a gentle breeze.

5.2 Active Solar

Active solar implies the use of a device to transform solar radiation into electricity or heat (Boxwell 2012). One common application is to use mirrors to focus solar radiation onto a series of pipes containing water. The Concentrated Solar Project in Kona Hawaii is located at the Hawaii Ocean Science Technology Park and administered by the Natural Energy Laboratory of Hawaii Authority (NELAH 2014). When the water boils, the steam turns a turbine. The turbine is connected to a shaft inside a copper coil. As the shaft turns, electrons are released, creating electricity.

A heliostat is a device that tracks the movement of the sun. Mirrors mounted on the heliostat focus sunlight on the tank of water. Temperatures inside the assembly can exceed 950 °C, incinerating unsuspecting animals which enter the heated zone. When the sun is partially obscured by clouds, an unrestricted heliostat can wander off track, thereby causing harm in unexpected places.

Mirrored surfaces can be used for cooking. For example, the Girl Scouts of America build solar ovens out of a cardboard box and aluminum foil. Commercial applications of this principle have been used to distill water. The City of Cincinnati Municipal Sewer District uses an Ultraviolet (UV) disinfection process to kill bacteria, viruses and other pathogens in drinking water (City of Cincinnati OH 2013; Cussler 2009).

Photochemical (PC) and photovoltaic (PV) devices convert light into an electrical charge (Häberlin 2011). On large-scale installations of solar panels in the United States, the resulting direct current is converted to alternating current and is transmitted along the electric grid. On a smaller scale, a single solar panel may be used to power a single electronic device. Excess electrical power in either instance can be stored in batteries for future consumption (Patel 2006).

Often solar panels are arranged in rows on the ground. The temperature directly below the panel is significantly cooler than ambient temperature. At a solar farm in Florida, the height of the panels was high enough to mow the grass beneath. The caretaker of the farm remarked that wildlife such as field mice and snakes sought shade beneath the panels, which in turn attracted birds of prey.

In countries with little arable land, solar-sharing farmers can generate electricity from the sun while growing crops. In Japan, Akira Nagashima developed a solar greenhouse to produce electricity and to grow plants. The solar panels in the roof reduce the amount of sunlight received by the plants by about 30 % without affecting plant growth (Takada and Watanabe 2014).

Installation of solar panels has inherent dangers. The most obvious is the risk of electrocution, because the panels are energized when exposed to light. For panels installed on existing structures, there is always a risk of falling from heights. The National Institute for Occupational Safety and Health (NIOSH) and the Occupational Safety and Health Administration (OSHA) recommend the use of a parapet around the perimeter of a roof to prevent falls. Solar panels are mounted on skids. During installation, these skids are a trip hazard. Wind can provide an additional challenge for installing solar panels. Panels should not be lifted by crane when wind speeds exceed 25 mph. While a steady wind at 6–10 mph may be welcomed for its cooling effects, a strong gust of wind can damage an unattached solar panel and may cause harm to persons or property nearby.

Active solar can be used with a heat exchanger. Solar hot water heaters absorb solar radiation to heat water. In one configuration, the heated water is circulated under the floors to heat the house during the winter. Heated water may be used to heat tanks of potable water or may be used in showers and appliances (DoE 2012).

5.3 Batteries

A battery is a device that is used to store electrical energy through a reversible electrochemical reaction. The most common batteries use lead and sulfuric acid. Other metals such as copper may be used. Electronic equipment such as solar panels, cell phones and laptop computers has increased the demand for light-weight, long-life batteries. The most common ion currently used is lithium. The largest lithium deposit is in Bolivia. There are known deposits of rare earth metals in China, the United States and Iceland. Unfortunately, the deposits in Iceland are contaminated by radioactive uranium deposits (USGS 2014). Mining the deposits of rare earth elements in the United States is hampered by environmental law. China has lax enforcement of their environmental laws. As the demand for and reliance on electronics increases, the geopolitical influence of China grows.

5.4 Additional Topics for Research

Heliostat, Sun room, Heat island effect, Sleeping porch, Veranda, Lake effect, Solar tracker, Parabolic troughs, Thermal energy storage, Organic-Based Photovoltaics, Organic semiconductors, Cadmium telluride tetrapods, Cadmium, copper, Wafer technology, Advanced Research Projects Agency-Energy (ARPA-E), Advanced Management and Protection Energy Storage Devices (AMPED); Batteries for Electrical Energy Storage in Transportation (BEEST); Grid-Scale Rampable Intermittent Dispatchable Storage (GRIDS); and Robust Affordable Next Generation Energy Storage Systems (RANGE).

5.5 Assignments

1. Determine the angle of incidence for your location at noon on June 21 and compare it to the angle of incidence on December 21. Then determine the angle of incidence for your location at noon on September 21 and compare it to the angle of incidence on March 21. Using these four angles, where would you place windows to maximize sunlight inside the building during the winter and minimize sunlight during the summer?
2. Compare the R values for brick, concrete, air, and wood. What is the recommended R value for a house? What material/s would be most economical to use in your area?
3. How many different seasons do you experience in your city? Would deciduous trees or evergreen trees make a difference in your home cooling costs?
4. Imagine you are building a multi-story house in a rural area. You want to have a safe, cool place to sleep outside during the summer. Where do you place your sleeping porch?
5. What is the American Institute of Architects vision for solar power in 2020 and beyond?
6. Compare and contrast the use of solar panels in nations with and without a national energy grid.
7. How is salt made? Compare the traditional method of making salt to a process that uses active solar energy to remove salt while distilling water.
8. The Toyota Prius 2014 has a solar panel embedded in the sun roof. How do Toyota dealers handle the replacement and disposal of inoperable solar panels? Is this a marketing trick or an ethical application of the technology? Did it increase sales?
9. Some communities recycle batteries. Some communities recycle computers. What happens to these two products when they are collected?
10. Lead-acid batteries are cheap in comparison to newer technology. These batteries contain lead plates, an acidic solution, and a plastic case. How can these components be recycled?
11. Where are lithium batteries used and how are they recycled?

References

Boxwell M (2012) Solar electricity handbook. Greenstream Publishing. ISBN-10: 1907670181; ISBN-13: 978-1907670183

Carmean K (Winter 2009) Points in time: assessing a fort ancient triangular projectile point typology, Southeastern Archaeology, p 2

City of Cincinnati OH (2013) Treating water with ultraviolet light http://www.cincinnati-oh.gov/water/about-greater-cincinnati-water-works/major-projects/uv-treatment-project/. Accessed 30 Sept 2014

Cussler EL (2009) Diffusion: mass transfer in fluid systems, 3rd edn. Cambridge University Press; ISBN-10: 0521871212; ISBN-13: 978-0521871211

Green D, Perry R (2007) Perry's chemical engineers' Handbook, 8th edn, McGraw-Hill Professional. ISBN-10: 0071422943, ISBN-13: 978-0071422949

Häberlin H (2011) Photovoltaics system design and practice. Wiley. ISBN-13: 978-1119992851; ISBN-10: 1119992850

Natural Energy Laboratory of Hawaii Authority (NELAH) (2014) http://nelha.hawaii.gov/. Accessed 30 Sept 2014

Patel MR (2006) Wind and solar power systems: design, analysis, and operation, 2nd edn, CRC Taylor & Francis. ISBN-10: 0-8493-1570-0, ISBN-13: 978-0-8493-1570-1

Takada A, Watanabe C (2014, May 27) Solar farmers in Japan to harvest electricity with crops. http://www.bloomberg.com/news/2014-05-26/solar-farmers-in-japan-to-harvest-electricity-with-crops.html

US Department of Energy (USDoE) (2012) Active solar heating. http://energy.gov/energysaver/articles/active-solar-heating

US Geological Survey (USGS) (2014) Lithium. http://minerals.usgs.gov/minerals/pubs/commodity/lithium/mcs-2014-lithi.pdf

Chapter 6
Water

Abstract Water is one of the most destructive forces on our planet, yet life cannot continue without it. Humans drink it, pollute it, store it, and harness its power. Plants uptake it through their roots and expire it through their leaves. Water is produced through combustion. As the most abundant greenhouse gas, it protects the earth from solar radiation.

Keywords Three Gorges Dam · Delta smelt · Panama canal · Endangered species act · Tennessee valley authority · Paraná river

Man has struggled for centuries to contain water, one of the most destructive forces on our planet. It is an elemental struggle. He places it in containers to consume later. He uses it to transport goods to market and for personal transportation. He builds obstacles to direct it into his fields and to use its power to grind grain. He builds reservoirs to store it and dams to generate electricity. He makes a covering over his dwelling to keep it out. Life cannot continue without it.

6.1 Currents

The flow of water responds to the pull of gravity from the sun, the moon and the rotation of the earth and to differences in chemical composition and temperature (Cussler 2009; Green and Perry 2007). Ocean currents movement in the x-y-z planes are also influenced by the depth of water, the proximity to land, and temperature variability with depth due to solar radiation, sun spots and solar flares (Philander and Holton 2003; Seinfeld and Pandis 2006). The prevailing ocean currents, shown in Fig. 6.1, influence inter continental trade routes (Garrison 2012). Rivers, streams and fluids in pipes flow downhill. The direction water flows and the speed at which it flows determines its suitability for transportation and industry applications.

P. E. Heckel, *The Ethics of Energy Sustainability*, SpringerBriefs in Environmental Science, DOI 10.1007/978-94-017-9701-6_6

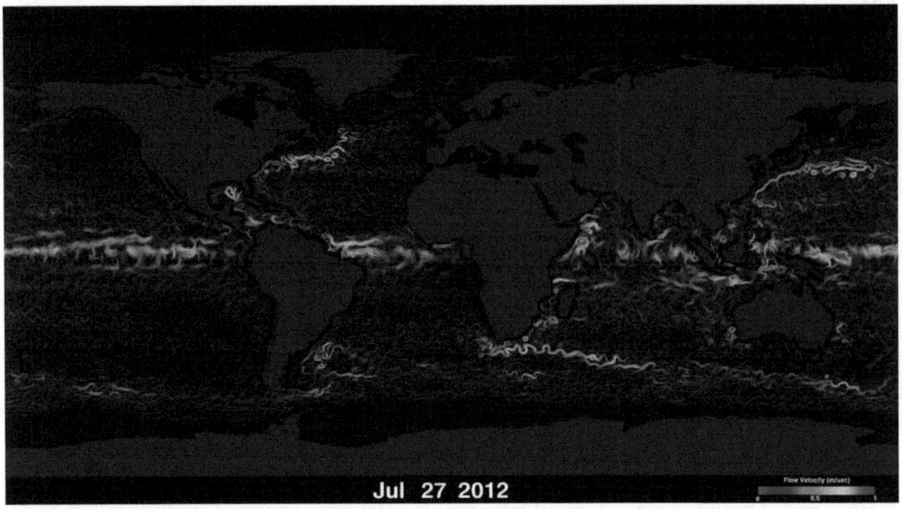

Fig. 6.1 Ocean surface currents. (Image courtesy of the National Oceanic and Atmospheric Administration)

6.2 Dams and Reservoirs

Ancient people found pools of water at various intervals along a stream. They may have observed beavers building dams from fresh cut trees. They may have noticed natural dams caused by storm debris, rock slides, or other natural phenomena. The fast flowing water was ideal for washing. The shallow water provided a place to ford the stream. The deeper water could be used for bathing. Over time, some species of edible marine life populated the pond. Thus through observation of the natural order of things, mankind learned to build dams and to exploit the advantages of still and fast water. The fast-flowing water on one side could be diverted into a chase to power a mill. It was easier to float goods across the still water on the deep pond.

Mankind experimented with building materials to harness the power of water. The first hydroelectric dam was built on the Fox River in Appleton WI in 1882 (America's Story 2014). The largest dam in the United States, the Hoover Dam was built during the Depression and dedicated in 1935. It provides flood control on the Colorado River; electricity and drinking water for cities in California, Arizona and Nevada; and irrigation and electricity for farms in the Imperial Valley. The Three Gorges Dam in China is currently the largest dam in the world. In addition to producing electricity, it was undertaken to improve commerce on the Yangtze River. The Itaipu Dam along the Paraná River border between Brazil and Paraguay is the largest operating hydroelectric facility.

The Tennessee Valley Authority (TVA) is owned by the United Stated government. Established in 1933 by an Act of Congress (18 CFR 1300–1399), its dual purposes are to manage natural resources and provide low-cost electricity (USG 2014). To meet peak demand, the TVA acquired land on top of Raccoon Mountain and built a pumped-storage facility. During off-peak hours, excess capacity is used to pump water into the reservoir on top of the mountain. During peak hours, the water is released through turbines to generate additional electricity.

Even a century ago, lawsuits were filed over every imaginable aspect of what was originally called Boulder Dam. This trend continues today. One of the first issues raised when dams are proposed in lands controlled by tribal councils is that dams hinder the migration patterns of fish. Another issue is the flooding of property behind the dam and the compensation of property owners under eminent domain. Plans to build dams and reservoirs in California are mired in lawsuits filed by environmentalists to protect a fish called the delta smelt (USDoI 2008).

6.3 Additional Topics for Research

Endangered Species Act, Water rights, Irrigation Canals, Arch-gravity dam, Masonry dam, Embankment dam, Barrage dam, Filled dam, Weir, Reservoir, Saddle dam, Cofferdam, Spillway, Fish ladder, Dyke, Diversion dam, Canal

6.4 Assignments

1. Prepare a report on the navigation system used on the Ohio River from Pittsburgh, PA to Cairo, IL.
2. Summarize the Rivers and Harbors Act of 1899 (33 U.S.C. 403).
3. Explain the purpose of an Intercostal Waterway. Give an example and show it on a map. On an annual basis, how many and what types of vessels use it? What is the economic impact? Delineate the differences between a port on the intercostal waterway and a port on the ocean.
4. Describe the Canal system originating in Augusta, GA. What was its purpose?
5. Explain the strategic importance of Peachtree Creek in Atlanta, GA. Which side of the Civil War conflict benefited?
6. Diffusion occurs when there is difference in concentration. Which ocean currents are influenced by diffusion? Which ionic species control the reaction?
7. Chemical dispersion refers to the motion of particles through a continuous medium. Start with a cloud condensation nucleus consisting of a particle of soot. Describe the formation of a single raindrop. What happens chemically when it falls into the ocean?
8. What is El Nino? How does El Nino affect weather in the United States?
9. How can sun spots and solar flares affect weather in Europe?

10. Describe the journey of a drop of water around the globe.
11. In March, 2014, in Washington State, a crack was discovered in Wanapum Dam, which lies on the Columbia River. The water level in the reservoir was lowered to allow inspection and eventual repair of the dam. Receding waterlines revealed old bones. The state Department of Archeology and Historic Preservation took possession of the human remains. Native people were notified of the discovery, but the tribal affiliation of the deceased was not known. Discuss the implications of this find.
12. Native people claim that dams have not only flooded ancestral sacred spaces but have also hindered the return of spawning Salmon. Each year, various groups petition the U.S. Army Corps of Engineers to remove dams. An environmental assessment is conducted, which includes a soils analysis for dioxins and other toxic compounds trapped through the silting process. Retaining the dam keeps these harmful substances from contaminating drinking water supplies downstream of the dam. Destroying the dam may open the waterway for fish migration. What toxins are of interest? What level of toxins is of concern? What is the decision point to permit the removal of a dam? Other than soil toxins, what other factors may be considered?
13. Identify and describe the parts of a hydroelectric dam.
14. What is the purpose of the Tennessee Valley Authority? What energy resources does it manage?
15. Research the Augusta, GA Canal built on the Savannah River in 1845. What were its benefits? How is it used today?

References

America's Story from America's Library (2014) The world's first hydroelectric power plant began operation September 30, 1882. http://www.americaslibrary.gov/jb/gilded/jb_gilded_hydro_1.html. Accessed 30 Sept 2014

Cussler EL (2009) Diffusion: mass transfer in fluid systems, 3rd edn. Cambridge University Press; ISBN-10: 0521871212; ISBN-13: 978-0521871211

Garrison TS (2012) Oceanography: an invitation to marine science, 8th edn. Cengage Learning. ISBN-13: 978-1111990848; ISBN-10: 1111990840

Green D, Perry R (2007) Perry's chemical engineers' Handbook, 8th edn, McGraw-Hill Professional. ISBN-10: 0071422943; ISBN-13: 978-0071422949

Philander GS, Holton JR (2003) El Nino, La Nina, and the Southern oscillation index. Reprint edition. Grand Central Publishing. ISBN-10: 0446679925; ISBN-13: 978-0446679923

Seinfeld JH, Pandis SN (2006) Atmospheric chemistry and physics: from air pollution to climate change, 2nd edn. Wiley-Interscience. ISBN-13: 978-0471720188; ISBN-10: 0471720186.

US Department of the Interior (USDoI) (2008) Formal endangered species act consultation on the proposed coordinated operations of the Central Valley Project (CVP) and State Water Project (SWP), 81420-2008-F-1481-5. http://www.fws.gov/sfbaydelta/documents/swp-cvp_ops_bo_12-15_final_ocr.pdf. Accessed 30 Sept 2014

US Government (USG) (2014) Title 18 conservation of power and water resources, Tennessee Valley Authority 18 CFR 1300–1399. http://www.ecfr.gov/cgi-bin/text-idx?SID=189deba1b62dc4760c190404a60722f3&tpl=/ecfrbrowse/Title18/18tab_02.tpl. Accessed 6 Oct 2014

Chapter 7
Wind

Abstract Wind in the tropospheric boundary layer is created primarily by thermodynamics. Wind in the stratosphere is generated by the rotation of the earth, west to east. Differences in chemical properties and surface characteristics affect the variability of wind patterns close to the surface. Understanding and mastering the wind has allowed us to transport goods and generate electricity.

Keywords Zeppelin · Windjammer · Climate change · Burbo Bank · Altamont · Kon Tiki

Wind presents us with another elemental challenge: how do we use it without being destroyed by it? Wind at the surface of the earth in the troposphere is created primarily by thermodynamics and the differences in various substances and surfaces to store solar radiation as heat (Cengel and Boyes 2014). The troposphere is neither uniformly thick nor homogeneous. Wind in the stratosphere above the boundary layer is generated by the rotation of the earth, west to east. Mixing occurs at the intersection of these two layers (Minto and Pleva et al. 2002).

Heat rises. Cooler air flows toward the heated surface. Water absorbs heat more slowly than land, so breezes blow from the water to the shore during the day. At night, the land cools faster than the water, so the wind blows away from shore, across the water (Green and Perry 2007). For centuries, humans used these conditions to navigate the globe. We experimented with sails and gears to grind grain and to pump water (Burton et al. 2011). We also studied the flight of birds, and discovered lift and drag.

7.1 Aviation

From the wings of Icarus to modern jet engines, man has been fascinated with flight. Hot air balloons, kites, zeppelins, airplanes, helicopters, sail planes, and drones, all have allowed man to soar with the birds. Momentarily, man conquered gravity to achieve flight.

Winged aircraft use the shape of the wing to create lift and drag. According to myth, Icarus used wax and feathers to create his wings. Helicopters use rotating

P. E. Heckel, *The Ethics of Energy Sustainability*, SpringerBriefs in Environmental Science, DOI 10.1007/978-94-017-9701-6_7

overhead blades to lift the aircraft vertically from the ground. A change in the pitch of the blades on the tail rotor allows lateral movement. Lift is also created when the force of the wind is distributed across the surface area of a kite. Kites can be flown when the wind speed exceeds 6 mph and are maneuvered by changes in tension on the string. The tail provides stability. Hot air balloons depend on the difference in gas density to create buoyancy, with hot air inside the balloon and cooler ambient air outside. The Hindenburg used hydrogen for buoyancy and diesel fuel in the engines for propulsion. The USS Akron used helium for buoyancy. Zeppelins use rudders to control lateral movement.

The type of propulsion mechanism determines the speed of aircraft and the economics of flight. A hot air balloon, sail planes, and kites are pushed by the wind. A discussion of powered flight can be found in the combustion section. Zeppelins use a small gas turbine engine for thrust and cruise between 20 and 40 mph, with top speeds approaching 50 mph. Airplanes powered by gas turbines and propellers travel faster than zeppelins, but slower than jets, which have a compressor section in front of the gas turbine. Cruising speeds for commercial jets approach 300 mph. Some military jets fly faster than the speed of sound. More fuel is used for faster flight.

7.2 Sailing

Sailing was invented numerous times by people living near bodies of water all over the globe. Mesopotamian Art from the Nile region circa 3000 BC depicts sailboats and other water crafts. Noah and his family survived the Great Flood of the Bible in the Ark, so surely sailing was a well-understood means of travel centuries before then. The sailboat allowed humans to travel to places far and near, to access the rich bounty of the sea, to conduct war and trade and to spread civilization. It was our earliest technology to harness the wind for profit and pleasure.

7.3 Wind Turbines

The modern wind turbine is a descendant of early sailing technology. More than a thousand years ago, the Dutch used sails attached to a wheel and shaft assembly to grind grain, hence the term windmill. The Dutch constructed a series of dykes and used the windmill to pump water from areas of the Rhine-Meuse-Scheldt delta (Smits 2001) The system of dykes and levees protected delicate estuaries and marshes from storms originating in the North Sea, Drier areas were cultivated. The windmill was adapted for industrial functions, such as a paper mill and a saw mill. Eventually, windmills were replaced by the more reliable steam engine.

There are several variations. Early windmills featured five or more sails. In contrast, modern windmills used to generate electricity have three composite blades connected to a central shaft and spaced at 120° angles. Tip speeds can reach 200 mph.

Windmills used to generate electricity are called wind turbines. To be economically viable, wind turbines should be placed in an area with constant wind, and access to the electrical power grid (USDoE 2014). Anemometers are used to measure wind speed. Tracking devices allow the turbines to rotate as the wind direction changes.

The power generating capacity of a wind turbine is determined by the size of the turbine and the wind speed through the rotor. Typical configurations can produce between 250 watts to 7 MW. China leads in the production of wind energy. One highly publicized wind farm in China uses 49 wind turbines to power 14,000 homes. An average size wind turbine, producing 3 MW, will generate enough electricity to power 1500 homes. In Mexico, farmers are paid rent for wind turbines on their farms; the contract may require them to report problems.

Meteorologists refer to the 300 m thick layer of air that lies close to the surface of the earth as the planetary boundary layer (PBL). A second air system exists in the portion of the troposphere above the PBL called the "free atmosphere". Violent storms may accompany the mixing of these two air systems (Minto and Pleva 2002; Seinfeld and Pandis 2006).

Friction and temperature differences between the air and natural features such as bodies of water, glaciers, forests, fields, mountains and valleys and manmade features such as pavement and buildings influence local wind speed and direction within the PBL. Solar radiation heats the land and water during the day. Differences in the physical properties of water and soil determine which retains more heat. This causes breezes to flow inland during the day. At night, the land loses heat faster than the water. Thus, cooler air over the land flows toward the warmer air over the water. This is a local effect. The wind speed at the edge of the beach is much different than the wind speed a kilometer away.

The rotation of the earth creates wind in the free atmosphere, above the PBL. The shape of the earth and the distribution of land mass determine the prevailing wind directions, with seasonal variations. At the poles, the wind blows mostly from the east. Across the Northern Hemisphere, to latitude 30 or so, the wind blows west to east, following to the rotation of the earth. Below that, the trade winds blow east to west. At the equator, the wind speed diminishes and becomes hard to predict. Across the Southern Hemisphere, the wind directions are reversed. A sailing ship would travel south to go east from Europe to North America and would travel north to make the return trip (see Fig. 7.1).

In general, the higher the elevation, the less variability there is in the wind direction. An ideal location for a wind turbine would be on a high mountain plateau; however other flat parcels will work. In the United States, wind turbines located on farms in Indiana have a reliable supply of wind due to the presence of the Great Lakes. Germany is placing a wind farm in the North Sea.

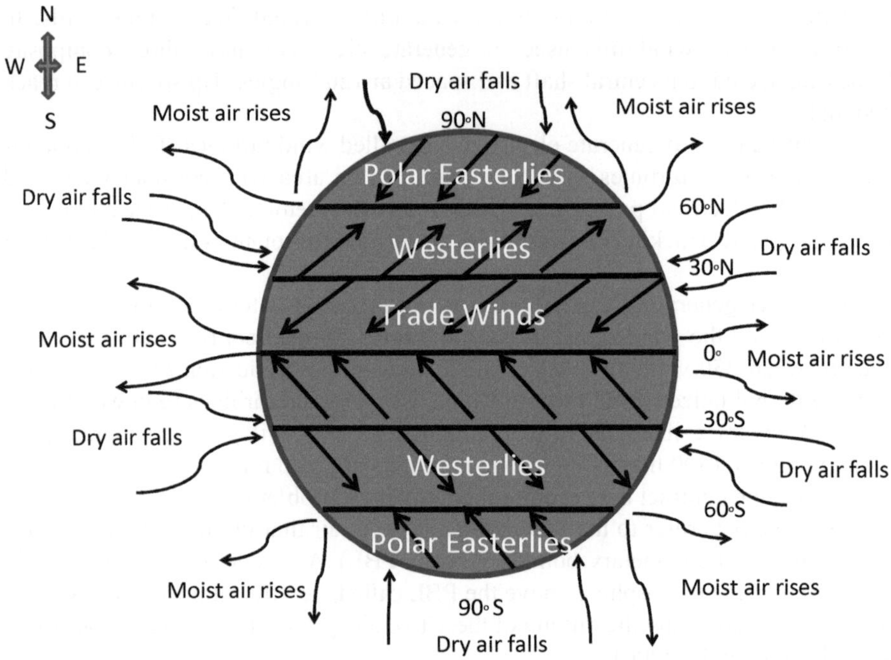

Fig. 7.1 Global Wind Patterns. (Image courtesy of P. E. Heckel)

7.4 Additional Topics for Research

Autogiro, Donald Cameron, Bertrand Piccard and Brian Jones, Steve Fossett, Richard Branson, Hindenburg, USS Akron, Igor Sikorsky, Solar sail, Electric sail, Magnetic sail, Windsurfing, Phoenicia, Leif Erikson, Vasco da Gama Christopher Columbus, Windjammer, Ferdinand Magellan, Kon Tiki, The Hudson Bay Trading Company, America's Cup, Jules Verne, Trade winds; Fuhrländer Wind Turbine Laasow; Mostert's Mill; Burbo Bank Offshore Wind Farm; Wind-powered water pumps; Altamont Pass, California wind farm, Polder.

7.5 Assignments

1. Who was the first person to sail around the world in a hot air balloon? How long did the trip take? What conditions were encountered? Was the voyage successful on the first attempt?
2. How does a helicopter fly?
3. What is the difference between the engines on a turboprop and a turbojet? Compare their fuel consumption rates and cruising speeds.

Wind Farm

Health & Safety	Environment	Economy
Pro:	Pro:	Pro:
Con:	Con:	Con:

Fig. 7.2 Wind farm

4. Wind turbine towers can be twenty stories high. There are no elevators. The tower has a design life of two or three decades, yet the turbine has a design life of five years. What are the health hazards for turbine maintenance workers? How can these be mitigated?
5. Sunlight passing between turbine blades can create a strobe light affect on nearby properties. When is this a problem and what can be done about it?
6. Many windmills are located in rural areas. Should public policy require the power generated by windmills to be used locally? Why or why not?
7. Identify and compare the 10 largest wind farms. http://www.windpowermonthly.com/10-biggest-turbines
8. CASE STUDY A wind farm is proposed in your town. Where could it be located? Use worksheet Fig. 7.2 to identify pros and cons of this project. Wind date for US cities is available from the National Oceanic and Atmospheric Administration http://www.ncdc.noaa.gov/sites/default/files/attachments/wind1996.pdf

References

Boyce MP (2011) Gas turbine engineering handbook, 4th edn. Butterworth-Heinemann; ISBN-13: 978-0814404720; ISBN-10: 0814404723

Burton T, Jenkins N, Sharpe D, Bossanyi E (2011) Wind energy handbook, 2nd edn. Wiley. ISBN-13: 978-0470699751; ISBN-10: 0470699752

Cengel Y, Boles M (2014) Thermodynamics: an engineering approach, 8th edn. McGraw-Hill. ISBN-10: 0073398179. ISBN-13: 978–0073398174

Garrison TS (2012) Oceanography: an invitation to marine science, 8th edn. Cengage Learning. ISBN-13: 978-1111990848; ISBN-10: 1111990840

Green D, Perry R (2007) Perry's chemical engineers' Handbook, 8th edn. McGraw-Hill Professional. ISBN-10: 0071422943; ISBN-13: 978-0071422949

Minto SL, Pleva E et al (2002) Planetary boundary layer; horizontal and vertical diffusion. http://www.met.tamu.edu/class/metr452/models/2001/PBLproject.html

National Oceanic and Atmospheric Administration. http://www.ncdc.noaa.gov/sites/default/files/attachments/wind1996.pdf

Patel MR (2006) Wind and solar power systems: design, analysis, and operation, 2nd edn, CRC Taylor & Francis. ISBN-10: 0-8493-1570-0, ISBN-13: 978-0-8493-1570-1

Philander SG, Holton JR (2003) El Nino, La Nina, and the Southern Oscillation Index. Reprint edition. Grand Central Publishing. ISBN-10: 0446679925; ISBN-13: 978-0446679923

Seinfeld JH, Pandis SN (2006) Atmospheric chemistry and physics: from air pollution to climate change, 2nd edn. Wiley-Interscience. ISBN-13: 978-0471720188; ISBN-10: 0471720186

Smits (2001) Windmills in Dutch history. http://www.let.rug.nl/polders/boekje/history.htm

US Department of Energy (US DoE) (2014) WINDExchange http://apps2.eere.energy.gov/wind/windexchange/

Chapter 8
Ethics

Abstract The fate of a nation depends on the morals and ethics of its elected officials. The practice of "quid pro quo", that is offering "this for that" generally refers to the exchange of services and/or money and/or political favors. When the responsibility for the enforcement of codes of conduct is assigned to a group of ethically challenged individuals, prosecution for ethics violations is apt to be tainted with favoritism, revenge, and political bias. The Ethics of Energy Sustainability revolve around deontological situations which must be confronted by persons of virtue who can use good judgment to resolve them.

Keywords Deontology · Virtue · Kant · Aristotle · Aquinas · Golden rule

Philosophers have discussed morality and ethics for centuries with very little impact on human nature. Western civilization in general and the United States in particular depend on ethical behavior to uphold the social contract (Wilcox and Theodore 1998). Laws are written to codify the social contract between people. Yet, ethics supercedes legality and transcends morality.

People behave according to their own moral code. A moral person will know right from wrong. An immoral person will not. Culture provides context for morality; however, cultural values may clash in a multi-cultural environment. For example, one culture may value hospitality while another culture values racial diversity. Is it racist or hospitable for a Caucasian to invite a Hispanic out to a Mexican restaurant? What if both are from Texas, but are currently living in Canada? Imagine other scenarios. Does your answer change?

Business ethics encompasses the assumed common morality between companies, shareholders, customers, and employees (Baggini and Fosl 2007). It is built on a foundation of trust reinforced by law. Corporations expect employees to share a common set of values which underlie morality and govern every day behavior. An employee who violates the company code of conduct is disciplined according to employer guidelines. This may include the loss of employment.

Shareholders expect corporations to behave in specific ways, too. When one corporation violates the public trust, shareholders react by selling their stock. When many corporations act improperly, Congress passes a law, such as the Sarbanes-Oxley Act of 2002 which codifies the fiduciary relationship between corporations and shareholders (USSEC 2002).

P. E. Heckel, *The Ethics of Energy Sustainability*, SpringerBriefs in Environmental Science, DOI 10.1007/978-94-017-9701-6_8

Customers have expectations. If these expectations are met, they will remain customers. A company's reputation is determined over time by the quality of goods and services it produces. Ideally, in the absence of a monopoly, a company which produces unsafe goods or defective products will be punished by the market and will go out of business.

Ethics is both contextual and situational. The social contract reflects the morality of the people. Sometimes values conflict. Thus, ethical behavior may require a choice between different cultural values. A recent plagiarism scandal is an example. Two researchers were working the same laboratory and often worked on the same project together. They came from a culture where collaboration was expected and the boss often took sole responsibility for work produced in the department. Researcher X asked to review the laboratory report of Researcher Y. Researcher X copied the entire report verbatim and submitted it to a journal as his own. The policies of the journal were that all significant contributors should be listed as authors. In this case, Researcher Y was the sole author and Researcher X stole his intellectual property.

There have been many philosophers who discussed moral behavior. Three of those are Aristotle, Immanuel Kant and Thomas Aquinas. Thomas Aquinas is known for the Golden Rule: do unto others as you would have them do unto you. This provides a relational context to ethics. The Aquinas principle helps individuals make choices between two values based on a relationship to those affected. Do you tell the truth or do you lie to protect a friend? Will the decision harm someone I know or benefit the community in which I live? Do my actions benefit the individual and harm my community? Are the effects mild or acute? Is this a one time thing or is it likely to happen again? Are the effects reversible or permanent? Do I enforce justice or show mercy? In essence, the Aquinas approach to ethics is relational. How do my actions affect others?

A moralistic approach to ethics evaluates problems objectively through the lens of right and wrong (Kidder 2003). These situations weigh actions against core values, principles or virtues. They may include some unfulfilled duty. They may violate a natural right. The response may not be entirely true. Actions may violate a professional code of conduct, a policy or procedure. Illegal activities may be involved. There may be really bad consequences.

There are three reason-based strategies to solve an ethical dilemma: consequentialism, deontology, and virtue (Anscombe 1958). Consequentialism asks who will be impacted by the decision and how do I measure the effects. It was developed by G.E.M. Anscombe to assert that actions with good outcomes were moral choices. Do the ends justify the means?

Aristotle pioneered virtue-based ethics. He was convinced that a person's character was formed over time. What kind of person am I? How do I want to be remembered? Aristotle advocated moderation in all things. Will my actions cultivate the virtue I want to possess? Do I know what the right choice is? Are my intentions and motivations appropriate for the action I plan to take?

A virtue based social contract evaluates actions based on how they appear to others. Would I be ashamed to read about it in the paper? Can I look in the mirror without guilt? How will I feel about this tomorrow? Would I want that to happen to me? Who in their right mind would do that? It may not be illegal but is it immoral? Is an achievement earned or it is based on power, privilege, or association? Could there be a conflict of interest?

Some virtues are valued universally and appear in many Codes of Conduct (Goldberg and Greenburg 1993). Do you have integrity? Are you honest and truthful? Are you candid? Are you accountable and responsible? Do you keep promises? Do you respect yourself and others? Are you trustworthy and loyal? Do you behave with dignity and courtesy? Are you kind and generous? Can you be fair and impartial? Do you discriminate against others? Do you have pride in your work? Are you patient? Do you keep your word? Can you be persistent when confronted by adversity? Do you understand your role in the organization? Do you know your limitations? Are you a life long learner? Do you strive for continuous improvement? Can you keep a secret? Are you competent? Will you act to protect the public? Can you make an independent decision or do you have a conflict of interest? Lapses in virtue can have lifelong consequences for yourself and others.

Deontology is a rules based approach which values actions motivated by duty (Cambridge 1996). Immanuel Kant's disciples follow certain principles they want others to adopt regardless of consequences. Right is always right. People of conscience are guided by reason. The individual cannot be sacrificed to save the whole, yet the individual can betray himself out of duty.

The areas of the social contract which are most likely to be controlled by rules are also covered by laws, specifications, and agreements. Reports and performance tests may be required to demonstrate compliance or competence. The moral temptation is often related to the data collected, the performance of certain tasks, the prevention of some behaviors, the guarantee of a specific outcome, or the assumption of risk. Delivery of a product, a guarantee of performance, and/or protection of the public may be adversely affected by lapses in deontological ethics. The Ethics of Energy Sustainability revolve around deontological situations which must be confronted by persons of virtue who can use good judgment to resolve them.

References

Anscombe GEM (1958) Modern moral philosophy. Philosophy 33(124):1–19. doi:10.1017/S0031819100037943

Baggini J, Fosl PS (2007) The ethics toolkit: a compendium of ethical concepts and methods. Malden: Blackwell. ISBN 978-1-4051-3230-5

Goldberg LA, Greenburg MR (1993) Ethical issues for industrial hygienists: survey results and suggestions. AIHA J (54):127–134

Cambridge texts in history of philosophy (1996) Kant: The Metaphysics of Morals. Cambridge University Press; 2nd edn ISBN-10: 0521566738; ISBN-13: 978-0521566735

Kidder RM (2003) How good people make tough choices: resolving the dilemmas of ethical living. New York: Quill, ISBN 0-688-17590-2

US Securities and Exchange Commission (USSEC) (2002) Sarbanes Oxley Act. Of 2002; Public Law 107–204—July 30, 2002. http://www.sec.gov/about/laws/soa2002.pdf

Wilcox JR, Theodore L (1998) Environmental and engineering ethics. Wiley. ISBN 0-471-29236-2

Index

P. E. Heckel, *The Ethics of Energy Sustainability,* SpringerBriefs in Environmental Science, DOI 10.1007/978-94-017-9701-6

MIX
Papier aus verantwortungsvollen Quellen
Paper from responsible sources
FSC® C105338

If you have any concerns about our products,
you can contact us on
ProductSafety@springernature.com

In case Publisher is established outside the EU,
the EU authorized representative is:
Springer Nature Customer Service Center GmbH
Europaplatz 3, 69115 Heidelberg, Germany

Printed by Libri Plureos GmbH
in Hamburg, Germany